PATIENT H69

PATIENT H69

THE STORY OF MY
SECOND SIGHT

Vanessa Potter

BLOOMSBURY
sigma

Bloomsbury Sigma
An imprint of Bloomsbury Publishing Plc

50 Bedford Square
London
WC1B 3DP
UK

1385 Broadway
New York
NY 10018
USA

www.bloomsbury.com

BLOOMSBURY and the Diana logo are trademarks of Bloomsbury Publishing Plc

First published 2017

British Library Cataloguing-in-Publication Data
A catalogue record for this book is available from the British Library.

Library of Congress Cataloguing-in-Publication data has been applied for.

ISBN (hardback) 978-1-4729-3610-3
ISBN (trade paperback) 978-1-4729-3611-0
ISBN (ebook) 978-1-4729-3613-4

2 4 6 8 10 9 7 5 3 1

Typeset in Bembo Std by Deanta Global Publishing Services, Chennai, India
Printed and bound in Great Britain by CPI Group (UK) Ltd, Croydon CR0 4YY

Bloomsbury Sigma, Book Twenty-four

To find out more about our authors and books visit www.bloomsbury.com.
Here you will find extracts, author interviews, details of forthcoming
events and the option to sign up for our newsletters.

For E, M and L

Contents

Foreword by Dr Tristan Bekinschtein

An accident is always a reminder of human frailty. We experience strong emotions when confronted with a surprising situation involving, for example, one minute going for a latte or cappuccino on a misty morning, and the next driving a loved one to hospital in an emergency. Brain accidents are particularly gruelling and emotionally exhausting as they affect not only the physical status of the person, but also their mind, personality and even identity. Whether such an accident is the product of a traffic accident or an internal brain event, you may not be the same if you recover, and your plans, dreams and life as you know it may be altered forever.

That is what happened to Vanessa Potter four years ago. She went from being a busy professional, married with two children and in good health, to becoming blind and losing her sense of touch almost completely within three days. In this book, Vanessa describes the progression of her illness in a gripping, engaging manner. Even though I knew the story I could not put down her text, and I still cannot let it go.

In the summer of 2014 I had returned from nine months of research in Australia, and was transitioning from neuroscientist at the Medical Research Council, Cognition and Brain Institute, to a new job as a lecturer at the Department of Psychology, University of Cambridge. Even though life was very hectic, I was relaxed. In fact, I did not find life in Cambridge very exciting in those days – I had nostalgia for my life in Oz, and the adventures and travelling. But then I met Vanessa Potter, a media producer who had willed herself to turn her life from that of a patient, to someone who made life an adventure for herself and others. She wanted to create a full-blown art-science exhibition, and for one of the installations she needed a scientist to help her turn her brainwaves into something that the audience could feel, touch and understand. This was where I came in.

Unlike other patients who experience an unusual (neuroscientists call it 'original') neurological event, Vanessa did not follow a traditional path to recovery. Her brain event terrified her at first, before she turned to analysis and learned to fight, reinventing the meaning of her illness and creating a new identity as Patient H69, which allowed her to grow. Throughout her illness and recovery, she began to find herself; using analyses of perception, memory, and a consciousness of her internal and external worlds, she went on to redesign and rediscover her self. This is a journey of recovery in the widest sense, far beyond just gaining a second sight. It is a trip into the soul and the brain that makes us truly understand who we are and what we need to change and, as Vanessa discovered, the fact that we need to use knowledge and our will as powers to aid recovery.

You think you know who you are, but actually it takes a whole lifetime to get to know yourself – unless you use the fast route and live furiously and take risks, but even that might not suffice if you do not come close to death or the full loss of your senses. The breakthroughs in life are scarce, and most of us do not explore them in enough depth to make them count for anything. It takes introspection to make the most of a significant experience – good or bad. This is what Vanessa did.

This is a beautiful book for many reasons, and I only mention a few that are dear to me. There's the personal, emotional aspect, and the depth of description of the essence and situation of the states of awareness that Patient H69 traversed. The text also goes into how people matter, and in what way those in your life matter. It delves into the core of what helping means, and what understanding is when a piece of the engine is broken. It takes you on the journey Vanessa goes on. The event, the loss, the slow recovery and the description of the characters creates the same visual stimulation that a play or film provokes. For a book about the loss of seeing, the scenery is fantastic. Then Vanessa confronts that difficult part of consciousness and its contents that we take for granted, and find hard to describe accurately. It is a place mostly reserved for poets and great novelists, yet we find it here, in the skilled writing of the story, the plot and the emotions intertwined with the science.

Lastly, we discover the core of what helping means and what understanding is when a piece of the engine is broken, and it is the importance of the people involved in this story that make it what it is. As a scientist and friend I am immensely proud of this book and of Vanessa. As a reader I am delighted to learn about, enjoy and experience the feelings, sensations and reactions she herself experienced. I struggled to write this foreword at first, but then I followed the path of Patient H69 – relearn, research, accept, go back to the fight, say what you think and feel, and describe what you have learned.

The Story of Patient H69

Let me introduce myself. I am patient H69. Of course, you know my given name, but for now I'm just Patient H69, because this story is simply about a patient, one of the many anonymous individuals, right now, lying in a crisp white hospital bed. We all have a story to tell – this just happens to be mine. Being a patient changed the contours of who I am. It has compelled me not just to tell my story, but also to investigate and uncover the science of what was happening to me on the inside.

When in 2012 I unexpectedly and abruptly lost my sight at the age of forty, the experience unlocked a deep reverence for the incredibly curative ability of my brain. Opening my eyes onto nothingness, my response was to take a deep breath in and dive down deep inside my inner watery self to find a way of mentally controlling my world instead. I sought an understanding and explanation of what was happening to me that was meaningful, and of some tangible use.

This psychological response was so primal that I was totally unaware of it at the time. Only much later, when listening to hours of audio recordings which I had made during that time, could I hear my own faraway voice narrating my story back to me. This has given me the unusual opportunity of actually hearing what I was saying during a time when all I can remember is the sound of my own blood rushing in my ears.

This book is split into two parts. Part One offers a diary-like account of what it was like to lose my sight over a period of just seventy-two hours. I then take some of the most peculiar and bizarre occurrences during my recovery, and try to bring to life the extraordinary visual rebirth I experienced. Here I start to ask questions and challenge myself with a series of visual tests that take me on a journey I had never imagined. This account is raw, and as far as I can manage a candid retelling of events, including the emotional effects my sudden blindness had on both my family and myself.

Part Two chronicles my own personal scientific voyage, during which I talked directly to scientists, unravelled my strange visual experiences and started to make sense of them. Here I encountered the difficult task of deciding which of the scientific revelations I had made should be included in this book, and I now know that this learning journey is not yet over – my illness was simply the start of it. I reveal the unexpected adventures and opportunities that fate had in store for me. The book draws to a close with how I now perceive a somewhat different world, and the tactics I employ to negotiate my new visual landscape.

I have tried to portray myself as neither victim nor survivor, simply telling this story how it was at the time, warts and all. There was an invisible army of friends and family who took over the reins to my life, and who kept a vigil by my side during one of the most harrowing periods of my life. They are present throughout this book, interwoven like silent guardians in among the story I tell.

Much of the content is written using the audio diary I recorded daily, so it openly portrays my response to the drama unfolding around me. I hope that this book not only gives some insights into my unusual visual reawakening, but also offers a glimpse of what some of the other anonymous patients lying in their hospital beds might be feeling right now.

This is the story of Patient H69.

Frozen moments

Telling a story is like painting a picture, so to build up the layers I offer a series of frozen moments snatched from my old life – the life I had before everything changed. Some of these memories catapult back to my childhood and to important events in my life, illustrating how the past, combined with our personality and upbringing, often sculpts the present.

We all have a survival instinct deep within us, and this innate response surfaces when we perceive danger, or find ourselves in a threatening situation. Every experience we have shapes our beliefs and attitudes, and affects the way in which we will respond – it paves the path we will walk, so to speak. Sometimes we can even pinpoint an isolated incident that happened years before, which caused a particular belief to emerge, and in turn created a set of behavioural patterns. These patterns can then have a huge impact on any future experiences, and indeed our relationships with others.

I had already developed strategies using visualisation to manage fear as best I could two years previously, so it's not surprising that these mechanisms were repeated again, even if I didn't realise this was happening. My original motive for using meditation was for self-preservation, but had I been asked at the time, I would have said it was to cope with the situation in hand. What I hadn't realised back then was that I was safeguarding the future, too.

For us to predict and understand our physical environment, our brains reference millions of past experiences. This constant feed of data is squirrelled away deep within the brain, and every piece of incoming information – every experience and every thought process – can affect how our brain is configured. This in turn affects how we respond and make decisions in the future, even when we're unaware of it.

Memories are a funny thing; they pop up sometimes, even when we don't know why. The memories I share were some of the ones circling inside my mind, hovering just out of reach during the onslaught of my illness. My brain was making connections, but at the time I didn't understand the significance of these recollections, sitting in the wing mirror of my consciousness. It's only now as I tell this story that I realise how important they are.

Permit me, then, to occasionally pause this story to share these vignettes with you, for they provide the otherwise missing context and meaning – the back-story, if you like. Collectively I hope they shine some light on *why* I responded to my illness in the way I did.

PART ONE
DIARY OF EVENTS

CHAPTER ONE

MMXII

M y story begins in 2012, when I was a seasoned broadcast producer making all those tiresome adverts you can now so helpfully fast forward through. I was happily married to Ed, whom I had known for more than ten years, and we lived in Crystal Palace, a leafy part of south-east London, with two small, energetic children. I should tell you straight off that Ed is the opposite of me, with a distinctly laissez-fair attitude towards life. He's the one swinging our children around by their arms and legs as I'm pushing him out of the door in the morning.

We were both, well, just normal. By the summer of that year I was at a bit of a crux in my life. Should I stick with producing, and spend 70 per cent of my time at work dealing with the many high-octane personalities that my industry attracts (that's the nice way of putting it), not to mention the increasingly unrealistic budgets and schedules that were being imposed on us more often than not?

In truth, my life involved sprinting out of work to squeeze myself onto an overcrowded train, my foot tapping impatiently at every red signal. I'd then jog to my car to don rally-driver gloves in order to pick up our children from their childminder in time. There wasn't a day that I got home without finding them fast asleep on the back seat when I finally turned off the engine. It was like living on the M1, and it didn't suit any of us. Ed had an arduous daily commute to a job he was less than enamoured with, so like many women who hide an invisible apron under their professional veneer, I was bearing the sole weight of our domestic life.

I therefore decided that as I was freelance (and the idea is that you can take time off), I would spend that summer with the children. No work, and no tantrums (I'm talking about adult ones as there is no escaping the juvenile variety). I was going to have picnics in the park with wasps and demands for the loo at the furthest point from them. We were going to have family fun, in quality time – *damn it*.

It actually turned out to be quite a stressful summer, as we didn't have a school place for our daughter (a stiflingly long story I won't go into here), so it wasn't all happy days. But at least I was getting more vitamin D than I had had in years, and I loved the lazy mornings and digging in the garden, even if it did mean that the house looked as if a tornado had just passed through it.

It was in September of that year, when our daughter was finally offered a place at our local school and the new term started, that a shadow started to creep over our lives. Halfway through that month, a flu-like virus hit me with crushing migraines and hot sweats that kept me bedridden for most of the time. I am not one for remembering specific dates – I tend to recall events by more visual means. However, I now have a date seared forever into my memory, for it delineates the path my life was to take.

On Sunday, 30 September 2012, I had recovered sufficiently for us to visit the Malvern Show in Worcestershire with family friends. I remember this date particularly, because it was to be the last day of that normal life.

Dragons and fairies, Saturday, 29 September 2012

iI'm sick of being sick. I've had enough of feeling as if I don't fit into my own body anymore. I am over-wearing sweat pants continuously for ten days straight, shivering and sweating at night, and am not able to face washing my hair. The whole smell of illness pervades me.

Today is the first day I can gently pull on jeans without my skin screaming abuse, and Ed smiles encouragingly as he hands me a cup of tea. With relief, I remember we are taking our children to their drama group, which is held in the back room of our local cafe. Ed and I gently push small bodies inside and close the door on animated howls and whoops that are no longer our concern.

Flopping onto a chair next to Ed, I feel I'm over the worst of the illness that has taken over my life. It's a delicate balance, and my wobbly legs move robotically, cautiously, every step an experiment. I'm okay, but only okay. I'd like to rest a bit more, but that just doesn't happen when you have two small bodies repeatedly bouncing on your bed at 6 a.m. I acknowledge with a silent sigh the inevitability that since I'm now able to stand upright again, I will simply jump back onto my life's roundabout and, one foot pounding the floor, get it spinning again.

Not uncommonly, we have convoluted family plans this weekend, involving travelling to the Malvern Show with my cousin Nick and his Spanish wife Maite. As I sip more tea in the sunny cafe, I find myself working out the adult-to-child ratio in my head. Thankfully, there will be four adults to our two children, which as I am acutely aware will ease the load a little. Ed and I have planned a night at his parents' house in Gloucestershire to break up the journey, so we're all piling into the car once our bouncing children are returned to us. I'm often surprised to uncover a nub of misgiving, a mild reticence at the thought of visiting my in-laws. This is not because they are anything other than reliably hospitable, but simply because they aren't my own family, and require just a little more conscious effort. There's always an unquestioned ease when you're with your own tribe, rather than an adopted one.

As I gaze out of the window, the bright sunlight illuminates an arc of sticky smears that little fingers have left behind. Staring absently, I am suddenly aware of an intense knotting in my stomach. My body is weak, and I feel bad-tempered and a little edgy. I normally love the Malvern Show, but this outing (which is always at my behest) is starting to loom over me now. Having a fun day out is all very well, but it's a different matter when you're not running on a full tank.

Malvern Show, Sunday, 30 September 2012

Maite and I have an indulgent half hour buying bargain seeds, having escaped the children and the menfolk. We peruse the

craft tents and shudder at the many crass offerings, novelty
peg bags and cheap bracelets that are guaranteed to explode
tiny beads across my living room floor in under a week.
However, there is one stand that catches my eye as I weave
past. Stopping and looking closer, I see they imprint children's
fingerprints onto silver pendants and cufflinks. Turning one
of the surprisingly heavy cufflinks around in my fingers I
make a mental note, and ask for a flyer, which I stash deep
inside my bag.

The men have given up trying to impress two bored
children with rides on tractors that they are not allowed to
touch. Instead, they have deposited them in a lopsided
children's fun-house with paint peeling off the walls. I can
hear the children scurrying around inside, like mice in a
cavity wall. The adults enjoy five minutes of contemplative
silence while nurturing polystyrene cups of milky hot
chocolate. Stretching up my arms, I turn a full 360 degrees
and breathe a sigh of relief. Flopping my head back onto
Ed's shoulder, I am content to no longer be locked into a
radar-like view of my children. The day has been frenetic
and has required much of my energy, and I am nearing
exhaustion.

Before long we are saying our goodbyes and offering each
other pacifying comments on how it will be easier next year
when the children are a year older, but I'm not sure if any of
us are convinced. Some local cider and huge sausage rolls
from the food hall on the way out have mellowed everyone
somewhat, and the day is considered a success.

The drive home takes much longer than we anticipated,
and the children are fractious and tired. I'm tired, too,
and find myself nodding off a few times while Ed stalwartly
negotiates the Sunday traffic. With military precision we
manoeuvre sleepy children into bed after donning pyjamas
without them noticing. By 9 p.m. I am utterly shattered. The
tuna sandwich I ate in the car has removed any vestige of an
appetite, so I head off to bed with a painkiller. My eyes ache,
and my body demands that I sleep.

TV static, Monday, 1 October 2012

I have that groggy, not quite woken-up feeling when the kids career past my open bedroom door, in search of a CBeebies fix downstairs. I am alone as Ed has already left for work, and I feel bleary-eyed and a little woozy. Gingerly levering myself up onto my elbows I blink a few times to clear my head, then try sitting up.

But the strange dizzy sensation is still there. Becoming mildly concerned, I slide out of bed to draw the curtains and click the light on to fast-forward the full waking up process. As the low-energy bulb slowly lights the room, I have to acknowledge that I am now as conscious as I will ever be. I am awake; yet I am still dazed. It's strange, too, but my eyeballs hurt and I feel a flicker of apprehension deep inside my stomach.

My mind whirrs with uncertainty, but I have to get the children to school and the childminder respectively – so I just have to *get on*. Switching into practical mode, I nudge my roundabout to get it moving again. Hesitating for only a beat, I call a new friend to see if she can take my daughter to school. My robot fingers punch the childminder's number, too, as I realise I can't drive safely with this level of dizziness. My mind is a beat behind my body and I am slightly flummoxed and unclear on the phone, but I manage to shift our normal routine. I then call my local doctor's surgery continuously until the phone is answered, and plead for an emergency appointment.

'What's wrong?' the receptionist's voice clips.

'I don't know,' I stutter, suddenly unsure of what to say. 'My eyes aren't right – I feel weird.' I pause, knowing I sound ridiculous, then blurt out, 'I just need to see a doctor!' It gets me an emergency appointment for 9 a.m. Ironically, I am aware that for once an early-morning call to the doctor is in fact an emergency.

My friend greets me as I open my front door with concern etched onto her face. 'What's wrong?' I am asked for the second time that morning. 'I thought you were better?'

'I don't know,' I mutter, 'I think it's my eyes. Something's not right.'

I see a shadow skim across her face as she steps back, gripping her little girl's hand.

'Keep me posted, won't you?' she whispers, guiding my daughter outside. The woman is genuinely worried for me, and her fear is infectious. As I close the door behind them the automatic smile that I slapped on this morning drops silently to the floor.

Standing unsteadily in our hallway, I am worried for myself. Shaking my head, I blink my eyes frantically trying to clear my vision. My thoughts are racing. I'm aware of an overwhelming desire to organise and articulate how I feel, but I flounder for the words to describe these indefinable sensations. I am catapulted back to a childhood memory of sitting huddled in our dentist's oversized chair, waiting nervously to have a tooth removed. I recall the curious sneaking numbness as the anaesthetic slowly crept over me, of smelling toffee and mint as my body melted, then oozed down a smooth, spiralling slide into unconsciousness. I have that same corkscrew feeling now, but without the childish anticipation of a lolly to wake up to.

Am I slowly losing consciousness? Is it even possible to lose consciousness slowly? Fuzzy static hums in front of my eyes with an alarming constancy; it won't go and I can't seem to shake myself free.

At 9 a.m. sharp I see a locum doctor who, to her credit, takes me seriously. I'm always concerned that whenever I see my doctor (normally with one of our children in tow) they will just dismiss me out of hand as a paranoid parent. I have to acknowledge that I'm not exactly walking in the door with a set of easily diagnosable symptoms here. *Feeling weird? That feeling when an anaesthetic kicks in?* She could easily smile patronisingly at me and put all of this down to being a stressed out mother who's just had flu. But she doesn't.

I am with the locum for more than forty minutes, during which time she thoroughly tests my vision and balance. One test where I have to follow her finger from side to side by only moving my eyes forces me sit down with a thump, overcome by nausea and dizziness.

At the sight of me with my head between my knees she calls A&E (Accident and Emergency) herself, muttering, 'I don't know what's wrong with you, but something is.' As she dials the hospital she turns back to me, her hand covering the mouthpiece, and whispers, 'You should call your husband.' Before I know it, a cab is here.

At this juncture I do decide to call Ed. I seem to call him several times a year with an emergency, but it's not normally about me. We agree to meet at A&E, which he doesn't question – but then he never says much on the phone.

At A&E I hand over the letter the locum doctor wrote for me, which I now realise I've had gripped in my hand the whole time. The receptionist smooths it out and smiles up sympathetically at me. This is definitely a less brusque reception than previous experiences here.

'It's my eyes,' I start unconvincingly. 'Something isn't right.' She nods and asks me to sit down and wait. Before long the reassuring silhouette of my husband wanders through the door and, leaning against him, I wait some more.

Eventually a bulky triage nurse assesses me, repeating my answers back in a thick Jamaican accent while avoiding any eye contact with us. After another endless wait, a very young junior doctor with a disturbingly trendy haircut finally arrives at our bay. He starts with the same tests the locum doctor performed, but can offer no suggestions as to what is causing my strange visual disorientation. I try to explain that I can see 'TV static', that I have a visible fuzziness right in front of my eyes, like the old-fashioned TVs used to have when the programmes had finished for the night. This description (aside from just ageing me) only seems to evoke a perplexed look from him. I am given an ECG, a chest X-ray is ordered and the doctor is gone as fleetingly as he arrived.

I try to describe my strange symptoms once again to Ed, but he shakes his head and whispers that I am not being clear, and that the doctors don't understand what 'TV static' means. It is evident that being stuck in A&E all day has caused him considerable stress. I know he's anxious to be back at work, and from the looks he's *not* giving me I start to wonder if he believes there's nothing wrong with me.

But I know there is.

Messages clearly aren't getting through. Why isn't there a universal word to describe this bizarre feeling? Maybe that is the problem; for what I feel is definitely more *sensation* than anything concrete, and that makes it all the more impossible to explain. Feelings are so infuriatingly subjective, and my unusually vivid descriptions are falling on deaf ears. I need someone to crawl inside my head and see what's going on for themselves – then maybe they could describe it.

A little more head scratching and more tests later reveal nothing conclusive; they cannot find anything wrong with me. By the end of the day I have the distinct feeling that they didn't believe me, or at least that they didn't understand me. Just how do you explain to a person whose life is medical, scientific and evidence based that you simply feel weird? In truth I haven't given them much to go on; it is only my instinct telling me that something is seriously wrong. The only obvious concern at this time is viral meningitis – the severe headaches and light sensitivity of the previous week could be pointing to that. Either way, I am ushered out of A&E and given anti-nausea tablets, which I never take because I know that isn't the real problem. As we near the door a convivial consultant drops in and says that if they weren't so busy they would keep me in for observations. I follow his gaze and see several old ladies with oozing head wounds queued up on trolleys in the corridor. This in itself is an odd sight, and I blink rapidly as I take in the scene. There is a strangeness to the world today; something is powering a deep sense of unease that I can't quite put my finger on. It's eerie to think what might have gone differently if I had indeed been allowed to stay in that night.

I go home straight to bed worried, yet quietly praying that I will wake up the next morning feeling normal again.

Spilt milk, Tuesday, 2 October 2012

I awake much more quickly this morning, alerted already by a clenching unease deep inside my gut. Almost before I am fully conscious, I have a strong sense of something amiss. Tentatively opening my eyes, I see that I am not registering the morning light as I normally would. It feels darker, with a brownish haze over everything, and even after blinking and rubbing my eyes it doesn't clear. Seeking an instant comparison, I mentally riffle back through my memory bank and realise that it feels not dissimilar to wearing dark sunglasses inside. Except, of course – I am not.

Sitting up quickly, my analytical brain assesses what vision I have left, and puts a number to it. I estimate that I have lost around 70 per cent of my sight, and the horror of that acknowledgement ricochets around my head like a metal bead in a pinball machine. I can still focus, and as I scan the room everything is where it should be, but it feels disorientating and alarming. It is all just *wrong, wrong…wrong*.

I hear Ed moving around downstairs. Unusually, he is at home this morning, as today is our daughter's fifth birthday. I can hear the muffled squeals of delight as wrapping paper is attacked and shredded in delight.

The smile drops off his face the instant he sees me venture cautiously into the room. We lock eyes, and our silent conversation confirms that we need to go back to A&E. No words pass as we cuddle and make a fuss of the excited little girl squirming around on our living room floor. She is absorbed in her special day so I manage to busy myself searching out the cupcakes I bought for her to take into school, and find and fill up her water bottle. I am distracting myself, maintaining the veneer of normal life, and it is remarkably effective. The morning routine is so embedded into my life that I can go through the motions without any thought. There is no time to assess my vision, to work out what I can or cannot see – I just have to do it.

As I pour milk onto the children's cereal, I notice that my left-middle fingertip is numb. I wiggle it about to make sure, but it is definitely frozen at the tip. A deep curiosity creeps over me. *When did that happen? Why?* Preoccupied, I drip milk onto the plastic tablecloth; but this morning I don't reach for the kitchen cloth to mop it up.

While the children are brushing their teeth with Ed, I find that I have floated my way back upstairs. Snapping into action, I dig out a faded brown rucksack with a broken zip, and start packing overnight clothes. I'm an expert bag packer, and even in my shocked state make sure I gather everything I might need for a stay in hospital. Gritting my teeth, I feel a grim determination – a steeliness – come over me. *This time they are not sending me home!* The voice is so strong that I wonder if I have spoken the words out loud. My familiar rucksack packed I sneak downstairs, keeping it out of view from the children so as not to spark a volley of questions.

Ed and I are both on autopilot, but hurrying out of the door we decide to deviate slightly from our standard plan, and swap the order in which we drop off the children. However, this simple idea is thwarted when we get snarled up in a rare queue of traffic just metres from home. Without thinking I ask, 'What's going on?' realising with alarm that I can't see for myself. Every stilted question that passes between Ed and I is loaded with fear and bewilderment. I try not to let my voice sound uneven, keeping the deep dread I feel internalised. I hold my breath and try to stop my rising panic from escaping and floating onto the back seat. Children are universally wired to shut out the loudest parental demands, but have an uncanny ability to pick up the smallest hint of something wrong.

Staring at Ed's profile, I silently will him to *know*, to really understand that I can't see properly. He has to know I am not being melodramatic. We are stationary for about ten minutes, the only noise being Ed's fingertips drumming rhythmically on the steering wheel. After winding down his window and leaning out, his strangely detached voice describes roadworks and a lorry stranded across the road. 'Is it foggy today?' I ask, as he clicks his seatbelt back in, but his clenched jaw answers my question immediately.

I realise I don't know what I should be seeing out of the car window anymore. It's as though I can't remember, but what I am seeing doesn't feel right at all. The roads are enveloped inside a grey blanket, and a huge fog has descended on the car. I force myself to remain calm, and relief floods through me when I recognise the street corner at which we are stationary. I don't know why that should be particularly reassuring, but I persuade myself it can't be so bad if I can recognise something outside. I am in a half dream. This place is somewhere I know; but at the same time it feels like an illusion.

As the traffic starts to move again I'm painfully aware that I'm commentating on everything I see outside, while Ed remains utterly silent. The more vocal I become, the more he sinks inside himself, and I can see his face slowly start to cave in.

My panic rises again when Ed drops me at A&E alone so that he can take our son on to the childminder. I am deeply unhappy about being left in the waiting area today. I see the same receptionist on duty, but I'm a little more forceful this time. My body is already adapting to my reduced sight and I instinctively lean towards her, insisting that I see the doctors as soon as possible. When she doesn't look up I hear my own voice reverberate around the room, quivering but loud, 'Please. I'm losing my sight!' and the sound of it startles me.

By the time Ed arrives back I am jittery and restless, only calming when his arm instinctively snakes around my back. Another nurse assesses me, infuriatingly writing down that my finger is painful. I silently berate her that my finger is not painful, only numb, and in fact it is only numb at the tip. My inner producer bristles at what feels like an important inaccuracy.

It is the same junior doctor who sees me again, but he still cannot find anything wrong with my eyes. We are becoming more insistent that something is done; there seems to be a lack of urgency, an apathy that hovers around my hospital bay today. The same genial consultant suggests that this time I am sent up to Ophthalmology, given that my sight is now physically affected. I am told that I can see a consultant ophthalmologist in his clinic that afternoon.

Finally, hours later, I'm deposited in a wheelchair, and as we pick up speed down one corridor I inadvertently grasp the armrest. As I do so I am vaguely aware that my right finger is now also numb. I notice this fact unceremoniously, as a fact, rather than as something to worry about. Tapping my left finger with my thumbnail, I feel that the numbness there has now spread further down. I am transported back in time to walking home from school on a freezing cold afternoon, trying to retract myself as far as possible inside the woolly depths of my duffle coat. I can feel the stinging numbness of my fingertips scrunched inside my sodden gloves. But I am not in the freezing cold right now, so while this numbness is familiar, it's frighteningly out of place.

It's now late afternoon and I am still waiting to see the eye doctor. Fidgeting like a child on the orange plastic chair, I am magnetically drawn to a white hospital notice board on the wall opposite me. Through my murky vision it glows like a beacon in the distance. When I came in I could still read the letters, but now as I glance back they are clouding at the edges, the words eroding from the outside in. I notice that all the punctuation marks have evaporated, as though a zealous cleaner has wiped away all of the black dots. I reflect on this pragmatically, almost without emotion. I have become a machine, a human computer noting these minutiae shifts and reporting them factually and systematically. I am reporting regularly, but I am not processing this information yet because it is way too big to take in. All I do know is that every blink is washing away more of my sight.

The ophthalmologist is a bespectacled Indian man in his late fifties, and his amiable smile instantly reminds me of Ben Kingsley's portrayal of Gandhi in the 1983 film of the same name. He calmly evaluates my sight, not making any judgements or comments. I can still make out the standard

Snellen eye chart he shows me, but I know this test is deceiving the doctors. I want to shout that all of the full stops have gone from the waiting area, but I know this won't help.

He hands me an Ishihara colour-chart book to look through,* and I'm thrilled when some pixelated numbers immediately dance in front of me. But as I turn the pages further along, there are others that are stubbornly hidden from view. Tipping the book on an angle, I try to somehow shake the numbers free. Smiling and taking back the book, *Mr Kingsley* doesn't look overly concerned at my strange behaviour; which in turn concerns me. Colour is my job, and I can see it draining away in front of me, yet this gentle doctor doesn't seem to understand this.

I try to explain that my sight is my job, that I know how to use my eyes better than most. I, of all people, would know if I was losing it, but he can't hear me. Finally, I am asked to do a visual field test that requires me to stick my head inside what looks like a large black plastic box. I am to press a buzzer when I see small pinpricks of light appear on the screen. I'm told this will test my peripheral vision, so I know I won't do well. As time has passed my visual world has narrowed, the outer details slowly but doggedly sliding out of reach. I now see through a halo, a brown circular border obscuring the outer world.

I'm right. The test results cause a frown to crease the affable face next to me. *Mr Kingsley* waves in a senior colleague, who instantly suggests that I need an fMRI.† As I stare at the backs

---

* Ishihara colour plates are made up of coloured dots and come in a small booklet. They are used by eye doctors to test for red-green colour deficiencies. These are pseudo-isochromatic plates, and were named after Dr Shinobu Ishihara (a professor at the University of Tokyo), who designed them in 1917. Nowadays I have a love-hate relationship with these annoying dots.

† Functional magnetic resonance imaging (fMRI) is a non-invasive technique often used in hospitals for measuring and mapping brain activity in patients.

of the doctors' bowed heads, I can see that the NHS (National Health Service) cogs are finally starting to grind into gear.

I'm pulled out of my reverie as I'm wheeled back into the waiting area, only to find the staff closing up the department. It is home time for the doctors, but I don't want them to leave until they have found out what is wrong with me!

A consultant ophthalmologist clicks past us on expensive heels, pausing only when she sees her colleagues huddled around my wheelchair. Attractive and business-like, she is a pastiche right out of a hospital drama. Flipping open her small leather diary, she suggests to Ed (not me) that if I don't improve I should see her a week on Friday. She is apparently *very* busy.

Hanging my head, I stare down at my scruffy trainers resting on the footrest. While I have been sitting here my toes have started to feel cold – really cold. *How did that happen?* Silently my heart thumps as I try to wiggle them, but they feel icy and numb.

I have no time to take in this extraordinary fact as yet another colleague is called to examine me. I had met this eye doctor before when I had a blocked tear duct a year ago. I was probably one of his more routine patients, so I am sure he doesn't remember me. As he hovers, his face just centimetres from mine, it strikes me that there is an out-of-place intimacy that you experience when being examined by an eye doctor. I am losing track of the number of pen torches that have been pointed at my pupils today.

The eye doctor is definitely taking more of an interest this time. He quickly asks the same questions as everyone else, but he's different – he really seems to *get it*. He is also very business-like, but he seems more convinced that he knows what is wrong, and he has a glint of what could be described as excitement in his eye.

Working swiftly, he performs a strange but (as it turns out) informative test. Asking me to look at him, he holds two boxes of paperclips in front of me, and asks which eye sees the box as the most red. It's a crude test but it obviously tells him what he needs to know. Speaking fast (I recall this habit from the last time), he announces authoritatively, 'I know what this is. It's going to get worse before it gets better; and it will take a long time to recover from. But you *will* get better.'

It's a bold statement in the absence of a scan, but I can retell his words verbatim, for that declaration began echoing inside my head from the moment he uttered it. Before he flies out of the door, *The Bold Doctor* tells me to come up and see him first thing in the morning. It's a very casual arrangement, but he knows I am to stay in overnight as I'm now serious enough to warrant an NHS bed.

I can tell Ed is tired as I am wheeled downstairs to a kind of holding ward while we wait for a bed. The dawning reality of my situation brings rolling waves of panic now. I stare at a high window, watching the evening light leisurely fade away. I know my constant questioning about what light there is worries Ed, and his unyielding silence silently shouts back at me. Clouds scurry across the unreachable prison window, and somehow I know that I am sucking up all this information in order to cling onto it – to store up these pictures for later.

My discomfort escalates sufficiently for me to be offered a sedative. I am apprehensive about taking anything, but am surprised and relieved at the effect. It does rub away the sharp edges of my fear, and it makes me feel calmer and somewhat distant. My body is lying on top of this bed, but my mind has levitated away and is floating high up above me. It is uncoupled and no longer attached to this world as I stare contentedly out of the window.

Eventually I am settled into a proper ward bed, and Ed has to go home. The original hirsute junior doctor tells me that I am to have a lumber puncture that evening so that they can test for viral meningitis. Two stilted attempts and plenty of anaesthetic later, and it is done. I even find my humour kicking in a little, and I make weak jokes in order to reassure the junior doctor and his nervous colleague. They both remind me of graduates I have recently trained, and I instinctively want to encourage them. As their footsteps fade away I find myself wondering if my jokes were for their benefit or my own, given the size of the needle they inserted into my back.

Handed back over to the ward nurse, I am instructed to lie flat for the next eight hours straight. Determinedly tucking in the hospital sheets around my legs, she announces that I am likely to develop a severe headache. Staring grimly at the commode and bedpan she tells me she has wheeled in next to me, I contemplate the incredible speed at which all of this has happened. Perhaps it is pride, but as I roll my head back to stare at the ceiling, I am determined not to use the bedpan *or* the damn commode.

Lying trussed up in my bed I realise with a jolt that not just two, but all of my fingers are ice cold – they have been anaesthetised. The deadness has crept up slyly on me, spreading ominously and silently from finger to finger. *Creeping.* That is exactly what is happening to me. Something is creeping inside my veins, slowly polluting my body, but I don't know what it is – and neither do they. My senses can't keep up with what is stealing over me. I tap each finger periodically to see if I can coax some life back into it, but can't feel anything. I have invisible Clingfilm wrapped tightly around each fingertip – you could prick one and it wouldn't hurt. It's an extremely uncomfortable and unnatural sensation. As I squint and hold my fingers up close, they look deceptively and frustratingly normal, even down to the bitten nails at the end. Yet right now they feel very, very different. In fact, they no longer feel like my own fingers at all. Everything is happening inside my head, and there is nothing to see. Only words can translate these bizarre and incomprehensible sensations, and I am not sure I have enough of them to do the job.

Even though night is drawing in, I am still reminding the ward nurses that I have to be at Ophthalmology first thing in the morning to see *The Bold Doctor*.

'No, it's not an official appointment,' I explain, getting more and more exasperated. 'He just told me to go.' They nod noncommittally and quietly take my blood pressure before they fade away again. I suspect I am not being taken seriously.

As the lights dim further, I start to crave human touch. I need to feel something solid, something *alive*. When anyone now approaches my bed I instinctively reach out. I seek a hand, a sleeve even. This obliterated sight is cruel, and I crave physical contact as a sensory replacement. I need to know names, too. I am building up an imaginary picture of all the nurses and staff from the taps and thuds of their shoes on the floor. A sharp tattoo signals urgency and a nurse unlikely to stop, but a hesitant squeak tells me someone is sauntering and perhaps already looking over at me. Together they create an invisible symphony, one I am quickly learning to interpret. Two days after my visual disturbances started, I am already developing strategies.

Faces are starting to disappear, and I feel a huge sense of loss. I miss the compassion and understanding that eye contact allows me. The nurses murmur sympathetically, but I know that they have no idea what is wrong with me. I have no label as yet, and that makes everyone who approaches my bedside particularly wary. It feels as if this is the calm before the storm; a building up of something – but none of us knows what.

Once the ward lights go out I coax my fingers to text Ed, anxious that he should meet me back here at 8 a.m. sharp. I have to keep shifting the angle of my head, straining to see any of the tiny screen. I am looking down a narrow passage of sight, one that is slipping away moment by moment. Forcing numb fingers to find the letters proves too difficult, so I use the side of my Blackberry case to press the keys instead. As I frantically punch instructions to Ed, I realise that like the notice board upstairs, my message has no full stops.

Rebelliously, I manage to hobble to the toilet unaided, but by now my toes have fused into frozen lumps. It makes walking cumbersome and nothing like the normal act it was a day ago. As my sleepless night progresses I am alarmed when my legs start shaking violently for seemingly no reason. As I obediently lie flat, desperate for sleep, it feels as though some malevolent force is taking over my body. My inner resolve grows in retaliation, as I try to control these frightening feelings. I resist help until a doctor is finally called and another sedative is prescribed. If I could clench my rubber-like hands into fists, I would.

Things move fast, Wednesday 3 October 2012

I am awake early.

Shit, it's got worse. More of my sight has eroded away – sleep has not cured me. Indeed, it's quite the opposite. I silently estimate that my sight has now diminished by as much as 90 per cent. That means I am seeing with only a sliver of my normal sight. Realisation punches me in the stomach – I am almost blind.

The truth of this horrifically calculated fact takes my breath away. The terror is so penetrating that it instantly ignites my organisational mode – that foot-stamping defiance that stabbed inside my chest in the night-time. I am fighting back, and as the galvanising waves wash over me I pull on my poker face. I know exactly what I am going to do, and there is no way I am lying down and giving in.

Sitting up I quickly pack my bag, using what limited touch I have left, noticing with surprise at how swiftly I am adapting. Without thinking, I use my elbows to nudge my belongings into my bag, as I can no longer grip with my fingers. In minutes I am ready to be taken up to *The Bold Doctor*. The nurses have stopped humouring me now and mindlessly repeat that I cannot go anywhere until the doctors have done their morning rounds after breakfast. Wouldn't you know it – my NHS breakfast is late today.

I start to get more vocal, insisting to anyone who will listen that I have to be let off the ward. *Where is Ed?* I mutter to myself. *Why does he always have to be bloody late?* Fear makes me irrationally angry that he isn't here, and my hands tremble as I clutch my bag.

Eventually Matron sees trouble brewing and silently materialises at my bedside. I can feel her starched authority hovering over me, but I lean over and grasp her hand anyway. I plead to be allowed off the ward. I plead with her because no one knows what is wrong with me, and the NHS isn't moving fast enough to save my sight. She silently fades away and I slump back crestfallen and lost. Minutes later I hear the hum of a wheelchair as Matron herself flicks back my sheets and eases me into the chair. No words are exchanged as we purr out of her ward, but I know we are moving quickly. Ghostly shapes merge into a grey fog as I hum by, and I sense a collective recoiling as my unseeing black eyes skim over theirs. I have no idea what Matron looks like or how old she is, but I know she is kind enough to break the rules for me this morning. This is a human being, not a member of staff helping me now.

I sit miserably in my wheelchair in Ophthalmology, still waiting for Ed. A young Asian nurse assesses my sight with another Snellen eye test. I can make out her waif-like shape, and her voice is whispery and youthful. To my dismay I can't see anything at all with my right eye now, and the left one can only register a tiny fraction out of the middle. All is black and murky around my peripheral vision, and I start to sob quietly. The young nurse takes my hand and tries to comfort me, 'You must pray. God knows best. What will be will be.' I don't agree with her sentiments – God isn't doing anything remotely helpful as far as I can tell.

Ed finally runs into the waiting area breathless and apologetic, and I furiously berate him for being late. My tears run unchecked now, and I know that other patients are staring at me as Ed wraps his arms around my shaking body. *The Bold Doctor* finally arrives and quickly evaluates the highly emotional woman camped outside his room. Ushering us inside he does some basic tests, but it's obvious I can barely see anything at all.

'OK,' he mutters. 'We need that MRI now.'

I can feel the NHS engine really beginning to gather speed; the unknown quantity of my condition is starting to make things happen. By this time Ed is consciously ignoring all of the hospital protocols and is not only wheeling me from department to department himself, but transporting my notes with us, too. Sod porters, they take too long and we don't have time – things are happening too fast now to follow the rules.

The Bold Doctor disappears to consult his senior colleagues, and I understand I have now made it to the head of the departmental food chain. *The Top Man* has a double-barrelled surname and is calm and composed – he's probably in his early fifties. I catch a glimpse of a striped tie and a subtle whiff of citrus. He reads my notes briefly, but then apologises and says that he likes to ask his own questions. I don't care as long as someone finds out what is wrong with me. We aren't in his room for long before he unravells his long legs and leans over to make a call. 'Fred?' he says amiably into the phone. 'I've got a lady here you need to see. Can she come down to Neurology now?'

In a blink we are sitting outside Dr Fred Schon's office. He nearly falls over my wheelchair when his door opens minutes later. Looking mildly surprised he nevertheless ushers me in politely. I can just make out a wiry, grey-haired man wearing what I think is a faded maroon jumper with a tie tucked in. In contrast to the snappy dressing upstairs, clothes are an after-thought to this man, and somehow that reassures me. His room is already full of clammy bodies, and we are introduced to several student doctors hovering like shadows in the background.

Briskly he asks me what is wrong, and it is clear that time and words are at a premium to Dr Schon. My reply is also economical: 'I can't see.' Ed starts to interject but Dr Schon cuts him off sharply.

'Thank you, but I want to hear from Mrs Potter herself, please.' Ed falls silent, and I feel his fingers brush mine. Complete authority surrounds Fred Schon – people listen to him and for the first time I feel safe. When he asks me to come through to his examination room, I realise with a penetrating horror that I can no longer walk at all. Arms fly in from each side to help me stand up, and as I wobble in front of him he asks me to roll back and forth on my feet. I can only manage an awkward stagger, which leaves me reeling in shock. *What has happened? When did it spread so far?* My feet are leaden, encased in freezing concrete – my body is being slowly frozen into stillness.

As I slump back into the safety of the wheelchair, I note that a patch on my left thigh is entirely numb, too. It hardly seems worth mentioning, given the severity of all my other symptoms. I am so shocked that I cannot take in these terrifying changes sweeping across my body. I look in Dr Schon's direction, unable to identify him among the swaying shapes. 'What's wrong with me?' I whisper. His voice comes back, slow and clear, 'I don't know yet, Vanessa, but I promise that I am going to do *all* I can to help you.' It is the first time he uses my first name, and he never calls me Mrs Potter again.

Yet again Ed himself wheels me back to the ward. My head is swimming now – I know I am seriously ill. Returning with a coffee and a cardboard sandwich sometime later, Ed resumes his position at my side. His large, warm hand is permanently holding mine now, silently transmitting our fear back and forth. His fear is different from mine, a low-level apprehension, masked by his unerringly calm response to any drama. The nurses have been to and fro taking blood from a variety of veins on my arms. I bruise easily and can feel the sore patches on my skin. Unable to look Ed in the eye, I squeeze his hand and tell him he needs to call people. I want my mum here now and I want my friends. We need a plan. Together we will make everything okay.

Ed is shooed out of the ward and I am left alone. The world is almost entirely brown now, like murky pond water, lifeless and dull. My unseeing eyes scan the room for anything remotely familiar to focus on, but the cold sweat that trickles down my back just confirms the futility of this. Looking down, I can still make out the numbers on my phone if I hold it right up, but my peripheral vision has all but disappeared. An eerie vignette has obscured almost the entire world around me. A shuffling sound makes me look up, and I sense a hobbling shape making its way unsteadily towards my bed. I feel panic start to rise; this place is surreal, overtaken by morphing shapes and sounds that I don't understand. A figure pauses by my side, yet I see nothing of the face that should be there. All humanness has gone, so when a bony claw touches my arm, I jolt back in terror. 'Don't worry, dear,' her emphysema breath dampens my skin. 'There's lots of help for blind folk these days. My friend Doris says they come to your house and everything.'

'Oh, my God!' My voice wobbles as I shake her hand off. 'I'm not going blind. I won't go blind!' I am terrified that this kind old lady believes I am permanently lost to this dark world. My heightened senses are overwhelmed by this simple gesture, this concern from another human being. Even though the evidence is starting to suggest otherwise, I will not accept that I am going to lose my sight entirely. I hang onto hope, onto any thread of light. I totally believe that my body will not give in to this.

Wheelie cases and kitten heels

Later on I'm moved up to another ward, which has around eight beds in it. For a ward with reasonably sick people, it is remarkably social. I feel a hand gently take mine and I am not fazed, even though I don't know who it belongs to. Jo is an outspoken South African woman inhabiting the bed next to me, and I am instantly drawn to her. I get her life story in a matter of minutes, and although I can't see her, I would guess she is in her late forties. I instantly like her, and understand she is the reason for the ward's friendly atmosphere. She whisperingly confides in me, giving me the ward gossip, and I'm a willing conspirator. Her vibrant descriptions and unstoppable banter colour my mind, and as she narrates the unfolding scenes in front of me, her voice becomes my eyes. I value her natural empathy, her unknowing ability to fill in my visual blanks. My black arms, slowly turning purple from all the blood draws, and my dishevelled appearance do not put her off, and I am intensely grateful for her company.

Quizzing Ed later he shrugs off my questions about Jo, telling me he thinks she has dark hair. I swallow my frustration at his vague replies. I suppose he believes these are silly unnecessary facts, but to me they provide texture and shape. These little details are keeping my ability to see alive, albeit inside the confines of my head. I need to practise seeing, even if there are only the vestiges of light passing through my eyes now. I need to hang onto what little I have left. I am an obsessive athlete exercising my ability to see, to make sense of the world, to engage with and understand it. The exact colour of Jo's hair, along with the cut, how she angles her head, her smile and the colour of her eyes are all extremely important factors to me. These are the things I would notice – if I could still see.

Anxiety is starting to creep in. A crushing sense of helplessness overwhelms me at times. Holding the bed rail I stumble towards the huge window next to my cubicle, and hold my hands out against it. The glass is cool, and as I strain to see outside I can feel the heat from the sun warm my face. I know the window is here because I can tell that it's the lightest part of the room; everything else is just dark now. I can feel the other patients watching me with a morbid fascination. I'm becoming aware of a palpable hush when my sight loss becomes obvious. I am vulnerable because they can see the horror on my face, yet I see nothing. My companions watch on curiously as on a bright autumnal day, I unknowingly stare directly into the sun.

Click, click, click; a pair of kitten heels is marching down the corridor. I'm instantly alerted to the accompanying whirr of mini-suitcase wheels. This person is walking with a sense of purpose and drive, and I instinctively know it is Mum. She quickly takes charge – medical emergencies are her thing, and she knows what to do. It's not visiting hours when she arrives, but all the rules are being broken when it comes to the woman in the corner bed. My mum moves in without much of a fuss. The case is put away in my bedside cupboard, and Matron is told in Mum's implacable but unassuming way that she is here to stay. Matron retreats quietly, knowing she is beaten by this diminutive but determined woman – this diminutive, determined mother.

Within a matter of hours Mum has taken control, not least of my rapidly unwinding household, but also of the nurses and staff on my ward. She makes demands, but not in a demanding way. I am relieved that the cavalry is here, and Ed can get some sleep.

My heartbeat is taking on a life of its own, thumping fretfully in my chest. I am uneasy, undone, and starting to feel my identity unravelling like a runaway ball of wool. There is an almost physical spiral of fear starting to pull me away from the people around me. The fear is deep now; it's a metal cage squeezing my heart tight, leaving me breathless. Without conscious thought I start to breathe slowly – in and out – in and out – the breaths getting longer and slower. Soon I am floating above my bed in a haze, in another place entirely. I am here, but I am not here. I can't see or feel my own body, so I don't know what is real anymore. My breaths get longer and more audible, and I can feel the guttural vibrations against my lips. I know I am exhaling so I can hear myself exist, so I know I am still alive. I am losing my understanding of my physical place in this world, but breathing is keeping me anchored. As I lie immobile in my bed, tears start to run silently down my face. I know I look awful, white faced and glassy eyed. I know that the abundance of jollity that everyone greets me with is simply a thin mask hiding their concern.

The teenage girl in the bed opposite me spots my tears. I know from Jo that she has learning difficulties and cerebral palsy. Suddenly her too-loud voice rings out, 'Me sorry you cry. Me sorry you cry.' I hate that I can't see her and I wish she would stop. I want to be invisible. I feel invisible.

Later on, after sending Ed home to sleep, Mum wedges herself into the hard visitor seat next to my bed and prepares for a sleepless night. Her instincts are correct as my leg tremors are on a whole new level tonight. I have no idea what time of the night it is, but I do know that my legs have been shaking uncontrollably for at least six hours now. I am exhausted to my core, and very, very frightened. The only respite I get is when Mum and I painstakingly circle the ward, hobbling around on feet now cocooned in three pairs of woollen socks – yet I still feel as though I am dragging fleshy lumps of ice. Under the thin NHS blanket my feet are entirely numb, not that the word numb really does this feeling justice. If I was a gangster who's fate was to end up wearing concrete shoes, I would know just how he felt. But this is no mafia film, and according to my brain, my feet really are encased in freezing cold cement. They are squeezed so tight that I am completely trapped.

Around dawn I start to sob, tired beyond any rational thought. I feel Mum's hot hand grip mine. 'I will bang on every door of every hospital to find out what is wrong with you,' she whispers fiercely. 'We will get through this.' Her voice breaks as she adds, 'I am not going anywhere.'

I believe her.

CHAPTER TWO

Snake Bite

Memory 1
A Boy, 2010

The tail of the dream wriggles away just out of reach as I open my eyes to find a face peering over me. Gasping in surprise, I flail around in the duvet.

'Sorry, sorry', Ed soothes me. 'I'm off, I was just seeing if you were awake.' Blearily I look at the clock and it flashes 06:04 back at me. In disbelief I turn back, but he has already gone, leaving the door slowly closing in his wake.

'Don't go,' I whisper into the empty air.

An hour later I am woken again, this time by a sharp, stabbing pain, and I know my waters have broken. I am overcome with a sense of conviction and utter certainty that my son will be born today.

Little fingers creep around the door, and my daughter's small nose peeks in.

'Hello, sweetheart,' I coo, not wanting to alarm her. 'Mummy is going to have a bath.' I know this is a mistake as soon as her face lights up. At two and a half years old, a bath with Mummy is a major treat. I know there is no way I will be allowed to bathe alone, so I lumber my huge frame out of bed.

The call to Ed is brief, with a 'told you so' thrown in for good measure. There are eight days before my due date, but there is now a huge blue birthing pool taking over our living room; the pool was blown up by Ed at midnight last night after a deep instinct demanded I had everything in place.

A text from my midwife tells me that she is at the dentist that morning, but she has passed my case on to another colleague in the homebirth team. She tells me she will get there as soon as she can. I have spent six months developing a relationship with this midwife following a terrifying experience during my daughter's birth. Yet, here I am on my own, no husband and no midwife due to arrive anytime soon. Breathing slowly and steadily, I lie immersed in a Hollywood-style bubble bath with my squirming, chatty daughter blowing soapsuds at my face. Even though I am getting regular contractions, a smile creeps onto my face. The funny thing is — now that it is all happening and I am here alone, no medic in sight, I find I really don't mind one bit. I don't have a secret midwifery team hiding in the cupboard ready to leap out and assist me; instead I have something much more powerful.

I have spent months of careful preparation — not preparation of my body so much, as my mind. The previous six months have seen me slide from a state of perpetual anxiety into what is almost a sense of detachment, a breezy acceptance of what will be. Every night for months, headphones on, I have walked through fields of startling blue cornflowers and gypsophila, with long grasses tickling my calves. I have told myself that this is going to be the most incredible experience of my life. I am entranced, hypnotised into believing that I no longer need fear — I simply need to breathe. As I lie in my bath my mind hovers above my huge, soapy body and is utterly content.

By the time the stand-in midwife arrives I am deep in my meditative state, calmly riding the rising waves that are sloshing over the sofa. It won't be long. I am vaguely aware of matter-of-fact voices, sighs from the midwife that she has been called too early and paper being shuffled, then a door bangs shut. Smiling inside, I know how wrong they are.

When Ed returns, my arm is already stretched out pointing towards the pool. His horror is silent — there's a realisation that the midwife has just left, and that all the hours of recordings he has seen me listening to late into the night are about to be put into practice.

My son roars into the world on a tidal wave of triumph, and it is a moment of pure and utter joy. The memory is instantly fused deep inside my brain. I realise I have untied the paralysing restraints of fear, and released a free-flowing euphoria instead. I am overcome by the miraculous ability of my body to know what to do. I am in awe of the ease in which my mind and

body simultaneously united with no inner dialogue or direction,
simply acting on a deeply seeded primal footprint.

In the warmth of the pool I hold my newborn son and
understand completely that I can change how I feel about any-
thing; that it is my choice. My drunken euphoria melts into
the droplets of water and drips silently into the swirling water
below.

Thursday, 4 October 2012

A random doctor stuck a needle in my leg this morning,
giving me a vitamin injection, and I didn't feel a thing. I
could sense his surprise but he said nothing. I am learning to
read these silent responses when things like that happen. The
medical community tells you when things are going well, but
clams up when it's bad news. The silences around my bed
are becoming longer and more loaded with every hour. The
nurses hover and whisper sympathetically; my leg tremors are
visible and painfully obvious, but no one knows why my
body is doing this. The numbness is far reaching, having now
snaked its way to the tops of both thighs. Parts of my back are
also feeling strange – not numb as such, but more as if it is no
longer my own skin holding this body together. I decide not
to tell anyone; they won't say anything anyway.

My consultant Fred arrives with his entourage, looking a
little flustered and out of breath.

'You were hard to track down,' he admonishes with a smile
in his voice. He performs some basic neurological tests, asking
what I can still see, and pushes his hand against the sole of my
foot to see if I can exert any pressure myself. As he returns to
the top of my bed I reach out to grip his hand, and although
I feel him tense a little, he doesn't remove it from mine.

'I'm going to move you to a specialist neurological ward
soon, and we're doing more tests. I'm also going to start a
treatment called plasma apheresis.' When he pauses, I hear my
own small voice, barely a whisper, ask, 'Is what I have life
threatening?' Fred's hand twitches, and the moment stretches
out endlessly in time. 'I don't think so,' he replies finally,

breaking the silence. It is the only answer I get, so it will have to do.

As Fred is leaving my mum sidles up to him and softly quizzes him, asking if what I have is Guillain-Barré syndrome. This is an autoimmune condition that she came across during her physiotherapy years, and the symptoms are similar to mine. However, Dr Schon appears mildly exasperated that a patient's mother is attempting diagnostics. He quickly puts an end to it and warns us of the danger of trying to compare my condition, whatever it is, to anything else. At the door he remarks tactfully, 'A snake bite is not the same as a bee sting.' I suspect this is an analogy he employs reasonably regularly.

I hobble around the bed exhausted, but with a strong sense of needing to keep moving. Unlike in my old life, that life I had only last week, food is now just a necessity. My stomach is permanently knotted but I know I have to eat, so I force down whatever limp option they give me at mealtimes. I now can't hold a fork myself, let alone identify the food I consume, so Ed gently feeds me. I can't help but imagine that the family looking after our two-year-old son at home is feeding him in the same way right now. As I chew slowly, Ed tells me what the kids have been up to, including the ongoing mishaps with our son's potty training. Some weeks ago, with no concept of timing, his four-year-old sister asked him if he wanted to wear pants instead of nappies. His blond toddler head simply nodded yes, and the decision was made. Our smallest child is fairly resolute, so when he decided no more nappies, that was what happened, pretty much instantaneously.

Ed reassures me that the kids are fine, and that they love having our family stay over. Our daughter is enjoying her new school, and is now in reception full time, thrilled to be wearing her red uniform. The children are asking where I am, which makes me cry, but I know I can't have them see me like this. I am undone, not myself, and need to stay strong. The touch of their hands would unravel me entirely.

To pass the time Ed has downloaded an audio book onto his phone. Taking my hand, he guides it towards the thin shape lying on top of my blanket. I cannot see any of the buttons, so he presses play, and loud voices instantly fill me up, replacing the sounds of the outside world. This quickly leaves me feeling peculiar, distant somehow. As I float out of reach I start to question my existence, because the everyday sensory cues that normally locate me have been suddenly cut off. Missing visual, tactile and now aural data I suddenly have no proof that I am here! In a wild panic, I pull out the earplugs and thrash around for someone to come to me, to touch me, to ground me once more. My senses are hauled back up to the surface by a scent I catch on the air, a familiar and musky smell, full of spice and warmth. I instantly know that it is Jackie. A curtain of dark hair brushes my face as kind arms creep around me. My friend is here, and we don't speak a word as she clutches my rubber hands.

Room 3, Friday, 5 October 2012

We finally get the confirmation that there is a bed for me at St George's Hospital in Tooting. At last the medical wheels are turning again, and we all agree that things just have to get better. A porter arrives to take me to the ambulance, but as I have Jackie's hand in a vice-like grip it's decided that she will ride with me. To be honest, it is non-negotiable as far as I am concerned.

I am beginning to understand that people observing illness from a distance often have a perspective that at times is entirely at odds with the patient. The nurses and porters prove to be a reliable source of gossip, and it soon transpires that I have been nicknamed 'the mystery patient'. I'm an enigma, a medical conundrum, and the case they all want to crack. But this is notoriety that I do not welcome. This afternoon's porter is more chatty than most.

'Come on, love, let's be getting you up, easy does it,' he chirps, but it takes two helpers to coax my alien body into the chair, and my right foot keeps flopping uncontrollably onto the floor. This awkward manoeuvring is not helped by the fact that I refuse to let go of Jackie's hand, yet no one seems to mind. The porter surprises me with his next quip, 'Bet you're wishing you bought a lottery ticket now, aren't you, love?' Jackie and I both go silent. 'What are you, a million-in-one case?' he continues. I stare mutely in front of me, trying to fathom what his comment means, and it suddenly dawns on me what he's getting at. I am overwhelmed by the insensitive, yet well-meaning complexities of human behaviour. The inexplicable fact that a porter is not only aware of my case, but has some statistics concerning it, is horrifyingly bizarre. It's so utterly inappropriate that I suddenly laugh out loud, and the return of humour relaxes us all for a moment. Laughing is what makes us survive.

I sense the cool afternoon air on my skin as the hospital doors whine open, and instantly feel vulnerable. I haven't been exposed to any natural light for nearly four days, and my skin is hypersensitive to these new vibrations. As the porter pauses, a gentle breeze winds its cool fingers around my bare ankles. I realise that I should have noticed a shift in light, however faint, but that hasn't happened. It's only now, sitting here in this wheelchair, that I can admit to myself that the worst has happened. The brightness of the day, the sunlight that I can feel warming my hands, has been sucked into a black hole, and I can see absolutely nothing at all.

I am intensely aware that everyone can see me, this hideous white-faced, dirty-haired freak. I pull Jackie down to whisper, 'Jax, it's all gone black,' and I can feel her silence, the enormity of what I have said rolling over her. *Why did we think I wouldn't go blind?* There was so little sight left; yet we were all grimly hanging onto it. Gripping her hand, I add, 'Please come in the ambulance with me.'

I can hear Jackie murmuring to the porter, and in a strained voice she collects herself. 'Of course, Ness, I'm going with you.' A further unintelligible conversation with the driver confirms yet again that another rule can be broken, and Jackie rides on the mechanical ramp up into the back of the ambulance with me.

I am now using my hearing as my guide, and ask a litany of mundane questions. *Is the door shutting now? Are you strapped in too? What equipment is in the ambulance?* Jackie's calm voice recounts everything, all the things I need to know – even introducing me to the elderly lady who is also being transferred to St George's. Jackie describes the traffic, the passing streets outside and how the sun is starting to set. I can't see any of it; the void in front of me is so black that it is suffocating. There is not a single shaft of light, no grey, no shimmers, nothing. I am intensely aware of my breathing, the noise of it, and the sensation of my chest rising and falling. I have a secret nagging fear deep inside me that even that might grind to a halt, too.

I am being taken to a *serious place*, obviously respected by the hospital staff that tag-team me there. This is the specialist neurological ward they had all been talking about. They lock the doors here. My wheelchair squeaks down yet another corridor to the end room – my own private refuge. The porter announces 'room three' with a flourish, as though I have just been wheeled into the bridal suite of a five-star hotel.

Ed swings open the door and, dropping my overnight bag on the floor, sinks down to hug me so tightly that I know he knows. Communication happens invisibly around me now; texts and phone calls swap updates on my situation in a continuous stream of beeps and buzzes.

My room has a wooden visitor's chair and a private bathroom, and Jackie and Ed describe it all to me in great detail. I am exhausted just swaying my way onto the thin bed, and I flop down motionless. Mentally transporting myself away from the confusion and noise, I breathe long breaths again and float myself to another part of my mind. I am not in this room anymore, but on a tropical island beach. I don't know when I first came to this imaginary haven, but it has been emerging slowly inside my mind. At the first wisps of panic an unconscious reflex makes me breathe my way gently to this colourful and safe place. As I allow myself to sink down inside myself, I recognise a loose familiarity to this place, and I start to understand that I can control the pace and ambience here. Room three is frightening and unknown, but I can be peaceful and calm here. I float inside a wooden cabin, looking out onto my deserted beach, and white muslin curtains flap gently in the warm breeze. It is early morning, a time before anyone else is awake, and the sun streams so vividly onto my face that I can almost feel it. Face down on my hospital bed, I stay on my beach as long as I am allowed to.

The rota for caring for me is being discussed, and I catch faint whispers as I drift along the sand. I am reminded that while my mind may be far away, my body is still here. My family has started to organise my home, recruiting helpers and taking over my life. My children are being taken care of by someone other than me. My life is being scrutinised and split into tasks. I think of some raspberry canes that I know are sitting in a plastic bag outside our front door, waiting to be planted out in the garden. I instinctively know that right now, everyone is just hurrying past them, not seeing them. Yet, as I lie here on this bed, I see them.

I am dragged back into the room by a loud, metallic squeaking as an oversized armchair is wheeled in on the sly. It is quickly nicknamed 'the marshmallow chair', as it is reminiscent of a squidgy 1980s airline seat that partially reclines.

Fred, or *The Big Man* as we are now affectionately calling him, is back. He arrives with a group of medical students and complements me on my new abode. His gaggle of silent scholars fidgets behind him as he explains the test results they have back. There are no diagnoses yet, but they still list some of the things I *don't* have. So far several possibilities have been mooted – multiple sclerosis and a brain tumour were two particularly grisly options, but neither hit the diagnostic jackpot.

It's hard being the patient in this situation; the first question I asked (and am still asking) is, 'What is wrong with me?' Rather naively, I never expected to not get an answer to that question. It is agreed that I have suffered some kind of neurological autoimmune disease, but that's where things are left for now.

Fred explains a little more about the course of plasma apheresis I am to start. During this treatment the overactive antibodies in my blood (which have been causing my nerve inflammation) will be filtered out, and replaced by nice new blood. This internal cleansing has a zealous ring to it, and sounds appealing. I feel polluted so I want the treatment to start immediately – I can't help but see it as a panacea, as some kind of miracle cure. I'm told that the clinician who manages this treatment has been asked to come in over the weekend especially for me, and I wonder absently if that has messed up his plans.

My attention is brought back when Fred, knowing it's my daughter's birthday party at the weekend, suggests that I be released for the day to go along. Shocked, I refuse to consider this suggestion for even one second. I want this blip to be temporary. If I leave this building, if I venture out into real life, my daughter would forever have the memory of her blind, wheelchair-bound mother at her party – and that is not a memory I want her to have. Inside the hospital my illness is routine. At a children's party it would be exposed and unnatural. It would give this horror show a reality it is not meant to have. I don't want that permanence, that branding upon my memory. Fred responds to my vehement refusal calmly.

'No problem, but you can go for the afternoon if you decide you still want to.' When he leaves I feel unnerved and agitated. Questions reverberate around my head. *Doesn't he realise what is happening to me?*

I discover later that there are a few practicalities that need to be addressed before my miraculous healing can take place. In order to access my blood for the apheresis, I need to have an intravenous line put into my neck. Nurses, increasing in seniority, have been attempting to do this most of the morning without much success. The more witty members of my ensemble quietly observe that my neck, with its erratic and gory puncture wounds, would make me a perfect extra in a vampire film.

I know I look awful. There is a sense of distaste that floats on the air when anyone approaches me. I wonder if my visitors fear that what I have might be catching; but bad luck isn't catching. Fear, however, is highly infectious. I can't see their reactions, of course, but in my over-heightened state I sense it all.

The phlebotomists have resorted to attacking the veins in my hands now to extract blood. My arms display a map of interconnecting black bruises, angrily shouting, *No more!* Ed winces as he gently touches the abused flesh, muttering, 'The kids can't see this, it would frighten them.' My stomach lurches at the thought of my appearance frightening my own children, and my resolve not to go to my daughter's party is reinforced.

I am methodically gathering visual information from those around me – building a mental picture of what I look like. I'm told my pupils are hugely and disturbingly dilated, the blue irises overtaken by black, unseeing orbs. I imagine that I resemble an owl, or with my increasingly unkempt hair, a zombie from a bad B-movie. I demand Ed takes a photo of me, just so that there is one. Either way, I know I don't look like the person I was five days ago; I am becoming unrecognisable. Something deep within me knows that I need to record all of this horror. I need to somehow hold onto all these rollercoaster feelings because I know that one day I will need to process them quietly, when this ride finally come to a stop. I ask Ed to jot down some events and names in a notebook, and I hear his legs cross and uncross in frustration. It's a strange request as far as he's concerned, and not the thing to be focusing on right now. But, while he's probably right, the narrator in me is determined to hold onto this supernatural world. I'm not quite sure why this is, but even I can hear the insistence in my voice.

Sometime later I find myself purring down the corridors being pushed towards Neurophysiology. I can't see any of the hospital so I only know where I am by the differentiating temperature or noise. The corridors are surprisingly draughty, and my squeaking wheels come to an abrupt and silent halt when they hit the rubbery lift floor. Soon I feel myself being angled into an airless room that I guess is small, as I instantly smell the odour of other people. Voices greet me, and I am clumsily manoeuvred into a cramped space. I feel my scalp being rubbed with something cold and wet, and a gentle male voice explains that he is putting electrodes onto my scalp so he can record my brain activity.

I'm then told a flashing light is in front of me, and to stare directly at it. I ask which direction the light is in, and the room falls silent. I guess then that the light must be pretty big, and is probably right in front of me. But, I have a black shroud suffocating me, so I see nothing. The silence accumulates and crawls along my arms and tickles my legs, and I can feel my panic clawing to get out. *I am blind!* The words bounce around inside my head. This isn't the silly throwaway comment I've made a hundred times before when the sun has got in my eyes. This is about to be definable, measurable – and real.

I'm dozing when my door swings open sometime later, and a nurse announces that I am to start a course of intravenous steroids and folic acid, along with the vitamins I am already getting. I also need to have blood-thinner injections daily. I blink at the long list. Needles jab me in the stomach and hip repeatedly, but I still feel nothing. The numbness is consuming me, and my clothes feel awkward against my skin. To my own touch, my body feels completely unreal, like a Barbie doll's, plastic and lifeless. This layer covering my body no longer belongs to me and I am uncomfortable wearing it. I try to nip myself with my deadened fingers, but this proves impossible. Even my nails, which for once are starting to grow longer, cannot provoke a response.

Apparently, I am in luck, the nurse tells me as he flips over the chart at the bottom of my bed, as I could have been stuck on an open ward. I raise my eyebrows again at yet another mention of the word *luck*. As far as I am concerned, it's all relative.

As the nurse leaves I once again catch the faint moaning that has been coming from a place on my left for most of the day. Whispers earlier confirmed that they are coming from a woman dying of brain cancer. I understand her anguish and confusion, but am mortified that I am also utterly devoid of compassion. My reaction to the nurse's comment about my private room now stirs a deep sense of shame. I have no room left inside me to care; my capacity for emotion is in the red already and I just want her to be quiet.

It doesn't take long for my blindness to totally envelop me. With no way to get my bearings, my sense of direction is now distinctly offbeat. Lying on the bed I have somehow sunk underwater, trapped beneath a black ocean with no concept of what is up or down. I am a diver submerged in the fathomless depths, and have lost all sense of direction. I have no reality. I sway slightly with the pull of the invisible current, and am reminded of the heavy resistance I felt as a child when trying to run in water.

I manage to drag my legs across the bed, and they drop like lead weights onto the floor. Holding the bed rail, I lean in the direction I believe to be the bathroom. Supportive hands immediately grip me and point me in the opposite direction. I am crushed and humiliated; yet it is only me who thinks that I should be able to orient myself. With horror I see that they know I cannot do it, and that I am putting an unrealistic pressure upon myself. Their silent sympathy is so real that I can taste it, and I feel more lost and more wretched than I have ever felt in my entire life.

Girl's night in

Today is to be a busy day. I am not allowed to rest as before long I am taken to have the promised fMRI scan. Thankfully, I know what this involves, but as I remove my wedding ring and the small silver toe ring I have worn for the last ten years, I feel remarkably naked. I plead for Ed to stay. Yet again rules are ignored, and he gently strokes my ankle as I lie listening to the clanking and whining of the immense machine.

Evening approaches and Jackie returns to stay over with me, tag teaming with Ed, so the next shift of family can take over at home. Apparently there are visiting hours here, but my family and friends creep around so politely that they are overlooked. I also require considerable help, so having keen outsiders taking over some of my care removes pressure from the nursing staff.

Jackie has lots of goodies with her, and I can smell them as soon as she enters the room. We have what I can only describe as a girly night in, which is pretty apt as it is Friday night, after all. She attempts to spray dry shampoo on my wild, matted hair, which is only partly successful. Balancing in my little bathroom I wash as well as I can with a zesty shower gel, and the acidic smell of lemons assaults my nostrils. I haven't showered in days, surviving only with these fumbled sluices. As I stand dripping on the tiled floor, I try to refamiliarise myself with an anaesthetised body I can no longer see.

Earlier in the day I'd asked Ed if I looked okay, and he'd simply replied, 'You aren't well, Ness.' I know I am staring with that off-centre blank gaze that blind people have. I know I look wrong, and Jackie is here trying to make me look right. She has also brought organic soft fruits and chocolate treats for us. I nibble the edges of an apricot and we distract ourselves with mindless chatter, reinventing a new normal. We gossip about the Jimmy Saville drama breaking in the news, a welcome albeit detestable story to take the focus away from me. This is a far cry from a Soho bar after work, but it's reality for me right now and we've decided to make my hell as comfortable as possible.

My daughter's party is playing on my mind, particularly as Ed is too busy looking after me to take over the organisation. Plus, arranging little girls' parties is not exactly his forte. In true producer form, Jackie and I discuss all the logistics so she can take over the arrangements. The small details I had put in place – the food, the cake, the pink plastic tablecloth, the games to play – are all inside my head. Jackie dutifully takes notes on it all, and I know the small people are in for a wonderful time. It is frustrating and sad to be missing the party, but I cannot bear anyone, including my own children, to see the monster I have become.

Eventually Jackie settles into the marshmallow chair to sleep, but any rest is interrupted by my gasps as spasms of pain course through my body. After various painkillers and sedatives (which Jackie and I weakly name 'train-crash pills'), I manage a couple of hours sleep. Even shifting my weight a tiny bit involves negotiating several cannulas and tubes sticking out of my body. As I lie still I feel totally medicalised, inhuman even. The world is dark and frightening, and I have absolutely no idea what is happening to me, or if it will ever end. These events, spanning less than a week, feel so timeless, so ingrained, that it would take years to shrug off this experience. Time is suspended; there is only this moment now, and nothing before or after it.

Apheresis, Saturday, 6 October 2012

Early on I am taken to a theatre to have an arterial line put into my neck, hopefully this time by someone who knows how to do it. This means that the magical apheresis can go ahead later that morning. Mum gently holds my hand while a young and intelligent voice dictates the proceedings. A tube is swiftly and professionally inserted deep into my artery. Well, I'm told it is, as I don't feel or see a thing. I ask the voice if he is northern, as I can detect a subtle twang. I also want to ask how tall he is, what he's wearing and what colour his eyes are, but I instinctively know this would be going too far. I long for a mental picture of this young man, but I know my questions would sound intrusive and strange. Society etiquette doesn't quite accommodate newly blinded women desperate for visual information.

Later I am wheeled off to start the blood exchange. I am clutching the hot-water bottle that I have begun to take everywhere with me, as my feet are still unbearably cold. I should point out that they *feel* cold, as in fact to anyone else's touch they are a frustratingly normal temperature. My brain can no longer communicate with my limbs, and my inner temperature gauge is spinning out of control.

I instantly like the apheresis technician when I meet him, even though he starts off by complaining about his journey into work on a weekend. From his accent I quickly work out that he is of oriental descent, and a kinetic energy surrounds him. I suspect there is a wiry leanness to his physicality, and I feel in safe hands.

I am hooked up to the apheresis machine and alarmingly it starts to bleep almost instantly. My blood isn't flowing strongly enough so, impossibly, I am instructed to relax. Unclenching my fists, I mentally will the machine to stop beeping so it can weave its magic and make me better. As the technician busies himself invisibly around me, he parrots his standard rhetoric. The only part I hear is that the treatment can take two to three weeks before I can expect any kind of results. This elbows my inner resolve, as every time anyone tells me something will take a long time, a voice inside me silently screams, *No, it bloody won't!*

Eating yet another stale NHS sandwich and sipping over-sweet hot chocolate, I wait for the treatment to be over. I hear the crinkling of plastic to one side, and the technician explains that he is removing the bad plasma and throwing it away. Mum describes the plasma as orange, but the technician just says that it looks bad. I am simply grateful that the misbehaving antibodies that did all this damage are being stuffed into plastic bags and burned.

As the sounds around me dim into the background, I reflect upon my inner outburst. I am agitated, my mind suddenly buzzing, thinking of all the things that need to be taken care of at home. I don't want to hand over my children's care – to trust those around me to do the right thing. I need to get better *now*. Surprisingly, thoughts of my children provoke an involuntary smile. The story of my small son proudly emptying the contents of his potty onto the kitchen floor while my dad was on duty has spread throughout my little gang of helpers. It's our funny story to repeat over and over again. It's something normal and amusing among all this scary stuff. The extent of Dad's confusion, and the size of the offending deposit, have morphed into unrealistic proportions to keep that flicker of humour alive. We need to laugh. My smile fades as I am suddenly overcome with a profound sadness, and I crave to be with my children again. I know that it's not just their tousled heads in the morning, or their cheeky grins I yearn for right now. It is the physical contact and that unchecked touch that children lavish on people they love; the neck-crushing hugs and kisses. As I lie here with the machine beeping angrily beside me, their absence is agonising.

Back in my room there is a treat in store for me. A cheery-voiced physiotherapist tells me that I am to get a walking stick. While this is a welcome mobility aid, I can feel my inner voice instantly object. Sighing, I know that it is a necessary evil, which will at least rein back the fingers that reach out to grip my elbows every time I slide off my bed.

Later, as I sit nervously on a plastic chair with wheels, Mum pushes me into the shower. The warm water is heaven on my strained body, and there is something womb-like about the experience. Water pools and squelches as I sit on the hard seat, but I feel rejuvenated afterwards. When we finally leave the bathroom, my hair wrapped in a hospital towel, we find a visitor waiting. My brother Dan has magically appeared and, never one to be shy about speaking his mind, instantly quips, 'Hey, Sis, you must be really sick to be given a room like this!' His inappropriate humour, so normal and familiar, evokes a weak smile from me, yet his quip resonates deeper than I care to admit. It doesn't take long before he has sloped off again when Jackie reappears for the night shift.

As the lights go out I slump back onto my bed and consider my situation. I'm as helpless as a baby; I cannot walk unaided and I can't see to do anything for myself. I have absolutely no idea where I am. Things couldn't really be much worse right now, so I figure it's time for a train-crash pill.

CHAPTER THREE

The Pillow

Memory 2
Bradford Girls' Grammar School, 1984

A piece of chalk skims frighteningly close to my ear, and bounces off the finger-marked wall next to me. My stomach lurches, as I know I have been spotted daydreaming again. Dragging my eyes away from the tempting window, I fidget and stare at the floor as I am reminded once more to concentrate. My accuser thinks I haven't taken in anything she has said. She is right, of course, but I have taken in other things. I notice the world, odd, sometimes random peculiarities, visual glitches. I know when the planet is slightly askew, when the natural balance I am so familiar with is upset. It might be a ruler balanced awkwardly on the edge of a desk, or the manner in which someone closes a door. Today it is the pattern the cherry blossom has made as it has settled on the grass outside. I've spent hours this week staring at the subtle pinks, the framing of the tree branches, how the light draws a line down one side of the grass. It touches me and transports me to another place. But, as a middle-of-the-road, thirteen-year-old student, 'noticing things' is simply considered daydreaming at school.

What my teacher doesn't realise, though, is that I don't just spot disturbances outside this classroom, I can spot them inside, too. My teacher was irritated this morning when she stood taking our class register, her mouth like a snapping turtle, eyes darting and accusing. The ring she had shown us all a month or so before flashed and glinted in the sunlight as she angrily ticked off our names. But what I noticed after lunch was not what was there, but what was missing. I was drawn to her eyes, slightly red, blinking erratically, with new sticky mascara. My gaze dropped to her ring finger, now soft, white and naked. To me it was a startling and obvious omission, a glaring sign — and as real as the delicate blossom on the grass outside.

Sunday, 7 October 2012

I stir; it's probably around 6 a.m. I seem to wake early, then dose until the obligatory Weetabix arrives after I've taken a train-crash pill. It's a peaceful time, and I float almost contentedly in this dream-like state. This morning, though, something is different. The black abyss didn't suck me in whole the second I opened my eyes. Still groggy, I begin to work out that the world is not *entirely* black; and that's new.

My heart begins to thud silently. To one side of me there is a pale grey, almost transparent shape I guess is a window. It is strangely two-dimensional, ghostlike and very surreal. Blinking, I struggle to interpret the visual messages my brain is receiving. Nothing is like anything I have ever *seen* before; but this isn't really *seeing*. This is something else entirely.

Moving my eyes down, I sense an even paler shape on my bed. Rolling onto my side I stare at this shape intensely for long minutes. My heart leaps — *yes, this is a pillow!* Staring harder, I can make out that it is roughly rectangular, and could those lines really be creases? I want to yell, I am so excited. I know what it is! Laughing, I punch and hug the pillow — this unforgiving, boring thing is now beautiful and amazing because it is the first thing I can register again. There are no other details for it is just a wispy shape, but a shape nonetheless. It is *there*; and if the pillow is there — then I am *here*.

Squeezing my eyes shut I open them in the direction I think my arms must be. Incredibly, two silvery shapes appear. *Is that an outline of my fingers?* I wiggle them and see movement. There's nothing else, no wrinkles, lines or anything to identify them as even human, but that doesn't matter. I know they are my arms. I exist; and a sense of my own humanity hits me like a gush of fresh air.

Jackie cries a little when I tell her, and her fingers are wet when she touches my hand. 'It's coming back. I know it's coming back,' I whisper over and over again. My face aches from all the smiling, and my inner child dances around the room.

I tell the nurses when they come in, and they flutter cheerfully around me. I try to explain exactly what the room appears like, but I have to be careful to avoid the word 'see', as although this is just a small word, right now it has over-sized meaning. However, I am starting to notice that those around me hear what they want to hear, not necessarily what I am saying.

How does anyone explain what he or she sees anyway? It's completely one-sided. This seeing that I am experiencing is nothing like anyone else's act of seeing. We are not even starting in the same place. My seeing is conscious, an effort and something I have to will myself to do, and it's exhausting. For everyone else it is an automatic reflex – a given.

As for what it looks like – well, this is another world entirely. I might as well have been transported to a moonlit monochromatic planet shrouded in mist – a planet where nothing is solid, and everything is composed of translucent flat shapes. There is no detail here, only confusing flimsy lines. This planet is so strange that I might have expected little green men to inhabit it, too, except that would mean there would have to be colour here; and there is none.

My belief is strong and I feel I can actually will my sight to return. I am suddenly validated that my body has been listening to me. Some primal instinct inside is commanding my body to heal and repair itself. I have to adapt pretty quickly to my new habitat, as nothing here is remotely familiar. Voices are the same, smells are the same, but the pictures don't match up. The atmosphere here is too thin, too watery, to reveal what I know should be in front of me.

I am starting to get some bearings of the space around me now. Darker shapes outline the edges of the door. I can tell when the main overhead strip light is on, but not the side lamps, and I can sometimes tell if the bathroom light has been left on when I hobble in myself. Over the course of the day I have silently worked out that this tell-tale behaviour belongs to my male entourage. If I could see the loo seat, I would probably spot it had been left up, too.

I toss words around my room, trying to explain what I see. X-ray vision, two-dimensional, a void, fragile – but these are not words you associate with seeing. Even though I can't see the expressions on everyone's faces, I know they don't get it. I settle on naming today a 'grain-of-sand day', the grave subtlety there for those that are listening. I have a beach to assemble, and today I have started with one grain of sand.

My brother is back to take over the familial shift, and his voice is welcome, even with the cynical undertones. As Dan shifts restlessly around the room he narrates the events of my daughter's fifth birthday party, held that morning at a local city farm. His voice ebbs and flows according to where he is in the room. From his account I gather that it was a noisy sea of princess dresses, interspersed with terrified rabbits being coaxed out of cages, only to be mauled by small, desperate hands. To my ears, that sounded like a complete success. The smooth flowing of the party appeared to be mainly due to the impressive efforts of his wife Marissa, and an abundance of bubble-gum pink party paraphernalia. I can hear him scratch his beard as he mumbles that it was his idea of hell. Good thing he's not a five-year-old girl, then.

Dan perks up when he starts telling me he has Googled one of the potential nasties still on the 'what have I got? list'. It's called NMO. He gets as far as quoting its rarity before I stop him. I refuse to be defined by three letters, an acronym. I can feel him bursting with the facts and figures he has hauled off the Internet, but I have decided that I am going to make my own recovery. The Internet cannot forecast what will happen to me, and I don't want to know about anyone else. My hand shoots into the air and I officially ban any other research.

Even chatting for half an hour has exhausted me, and I have shooting pains running silently up and down my legs. As I close my eyes, pictures and sounds ricochet around my head, bumping and jostling each other. I want to zone out but not to sleep. I just want to shut out the world for a while so my brain can rest; it's a war zone in here.

Sometime later, I can tell from the intermittent wafts that today's culinary offering is curry. I am not enthralled, as it smells distinctly bogus with the overly sweet, powdery smell of mass-produced curry powder. Holding a laden fork awkwardly, my brother attempts to wrap my bulbous fingers around it. I use a pincer technique, as it is the only way I can grip the metal. Our jokes are feeble but I can't bear my fingertips to touch anything. I ignore the awful sensations and deliberately feed myself, occasionally missing my mouth altogether. It is unbelievably difficult and the effort leaves me exhausted again.

My brother busies himself, quietly cleaning up the food mess on my bed. I smile inwardly as I am aware that this fussing would amuse his wife enormously. Dan is not known for clearing up his children's mess at home.

As Dan brushes away the crumbs, I start to flex my feet. I have been attempting some clumsy Pilates exercises in the last few days while lying on my bed. Obviously I have no idea if exercise is the right thing to do, but I have a strong sense that I need to keep my blood flowing. I fear that if I stop moving my whole body might be permanently immobilised. Movement is cathartic – even just tensing my muscles minutely reassures me. My legs don't belong to me, but I heave these lumps of rubber around the bed anyway.

Barbeque, Monday, 8 October 2012

'You could have a great barbie out there,' my brother yawns as he pulls up the window blind, letting the early-morning sunshine stream in. He stretches noisily and I hear comic squelches as he attempts to extricate his sleep-crumpled body from the marshmallow chair. His fidgeting last night suggested a restless sleep, but I am still hovering in between consciousness and train-crash sleep myself, so I gently bat away his random chatter.

Mentally scanning my body, my feet still feel painfully cold, but there's also a tightness that makes me wonder if someone has wrapped them up in gaffer tape too. I shrug off this ridiculous notion; my brain is still playing tricks on me. I sense Dan shuffle over to my bedside, his silhouette ruffled and dishevelled with crooked edges. His movement gently wafts sleepy smells over me and, still groggy, I reach out and touch his rumpled pyjama bottoms. Since my sight spluttered back into life yesterday, I am being pulled towards those patterns and shapes that float around me.

'Are those mine?' I ask him, but Dan doesn't answer as he's already lumbering off in search of the nurse's desk, no doubt in boyish hope of a morning coffee.

He returns moments later. 'They laughed at me,' he sulks, his voice sheepish. 'Hardly surprising if you will insist on wearing *my* pink pyjama bottoms,' I snigger. 'It was all I could find...' he shrugs. I know this is very likely to be true, as my brother has a limited interest in clothing, often leaving his cast-off clothes littered around his own home, on trains and at our house.

As I start to shift in the bed I feel incredibly stiff. This is becoming a common occurrence on waking up. Dan hovers protectively when I start to slide off the bed, worried that I might crumple to the floor. I know I am being bloody-minded, but I still refuse any help; my independence is too valuable to give up. My right foot is dragging on the floor as I shuffle along, but this morning I don't care.

Dad arrives early to take over from Dan, and I am left to my own toilette this morning. My father lives in South Africa normally, and only spends summers in Britain. After three months here he is due to fly home very soon. My illness, unexpected for all of us, is causing him some distress as he would rather stay on and help.

Lying on my bed having only dared a cursory wash in the sink, I am worn out again. Slowly manoeuvring myself into the bathroom, even with my newly acquired stick took a huge effort. Pushing myself back up onto my elbows I ask Dad what he can see out of the window. Sighing and shutting

Figure 1 Room 3. Credit: Photography by Ed Potter

his laptop lid, he walks over and looks out, 'Well, it's a bit strange, actually. There's this huge balcony with railings all around it, and buildings in the distance.'

'Ahh,' I smile to the ceiling. 'Perfect for a barbeque, then.' I can't see Dad's face, but I can sense his puzzlement.

Some time later my dad notices me clutching a plastic Evian bottle, staring fixedly at the label. I know my behaviour is starting to become disturbing, but no one seems to understand. Over the last few hours I have begun to identify the shadowy patches on the bottle – *as letters*. Dark silhouettes float eerily against a pale background when I hold the bottle centimetres from my face. Not understanding what my version of seeing is, they all ask me what the letters are, and I have to waft their nonsensical questions away. This isn't about *seeing* like they *see*; this is about sensing that something is *there*. I am using different parts of my brain to register this information; other inexplicable biological factors are at play here, even if I don't understand them. I try and tell them I'm not at the seeing stage yet, but they just scratch their heads. I go back to staring. Something is there on this squidgy plastic. I can't read it yet – but I know it is there.

There are shapes and letters emerging in the bathroom, too. I lean so close to the soap dispenser that my nose touches it. I know there is a short word there, but it's no more than a dirty smudge about four centimetres across. I promise myself that I will look at it every hour, every day, until I can read it. I shift my gaze to the unseeing mirror, gently touch my face and mutter, 'Hello.' The bathroom is my secret place where I can talk to the ghost I have become. I need to touch my face and remember what I look like, so I can make myself reappear. I stare down at my hands and silvery outlines move like my hands move – and right now I think that is bloody brilliant.

Frustration has me taking my glasses on and off a hundred times over the course of today. Ironically my vision is sometimes less confusing without them. Confusing is a word I am starting to employ frequently. I seem to have only some elements of vision reappearing, and I start to wonder if my sight is returning in layers. If so, this layer only has texture, but no sharpness, no contrast and no colour – so I guess I will have to wait for those.

My room has been transformed into a living X-ray, and grey lines flicker and shudder uncontrollably. Each blink takes a picture, a flimsy outline that provides limited information. It is disorientating and I feel there should be a sign saying 'Under Construction' in the corner of my eye.

Earlier today the brain-cancer lady next door started yelling, 'Bugger off!' over and over again. It might have been comical except that her voice sounded unnaturally thick, and unearthly. We listened in silence, each of us willing her to stop, but no one daring to say so out loud.

The chirpy but shy physiotherapist with the ponytail is back, this time with an occupational therapist friend. Their manner is pleasant, and their bubbly enthusiasm trickles over me. I am particularly engaged with their banter right now, as they have brought a test for me. Passing a number of objects to touch, I am asked to respond with whether I think they are soft, hard, scratchy or whatever. Of course I realise they are testing my sensory abilities. Even though my fingers are still numb, I am elated that I can recognise a ribbed hot-water bottle, a glass jar and a latex glove. They make me stand, sit and walk, then repeat it all over again. They pinch my bottom and tell me to tense my gluteus muscles, but we are too busy for jokes.

They leave with friendly but firm instructions not to hold the bed frame to balance, and to stand up fully. When they leave my body feels unbalanced and jelly-like. I can't seem to remember how to walk properly, and my legs flick out in front of me rather than stride smoothly. The girls have instructed me how to move my hips, and how to swing my legs. They have just coached me in how to walk, and I am shocked to realise that I can't automatically remember how to do that.

I am fired up, so when Marissa, my sister-in-law, arrives to stay over, we both respond to my homework with gusto. Within minutes she is on the phone to Ed demanding that he bring in sensory items from home. I listen as she tells him that my brain needs to respond and remember what textures *feel* like. It seems crazy, but I will do anything they tell me to do. I will wiggle my hips the right way, I will tense my bum and I will stroke any object they put in front of me if it will make me better again. As soon as I can face it, I am doing those exercises again.

We have the radio on almost constantly now, so I am more up to date with current affairs than ever before. For variety, Marissa and I try tuning into a radio play, but within moments I hear her scramble back up to turn it off. *Dracula* is not the best choice of drama at the moment.

A new doctor arrives and informs us that he works with *The Big Man*. He is obviously young and yet has an air of authority about him, and we playfully nickname him *The Protégé*. He does the same tests as everyone else, conscientiously tapping the soles of my feet and running keys down my toes, while asking if I can feel anything. Mostly I can't. I am conscious that neurologists don't tend to carry stethoscopes around their necks, as it appears their most useful tool is their front-door key. I have become an expert in recognising the hierarchy of those who push open my door by their voices and steps, or in some cases, unsure shuffle. All these audio cues (ones a sighted person can easily ignore) are now sensory gold dust to me.

Mum arrives for the night shift and Marissa goes home to Dan. I try a last walk to the bathroom using my newly tensed up bottom muscles, and have to admit, it is a tiny bit better. I am more stable. Mum scrutinises me from afar, her head cocked to one side. I am standing unaided – and that is progress.

Fashion statement, Tuesday, 9 October 2012

It's 2 a.m. and I need the bathroom. Mum turns on the toilet light and the room is illuminated like a premier-club football pitch. 'Wow,' I exclaim. 'I can really see that. It's light, really light.' It doesn't matter if it's the middle of the night; we grin like children.

By 8 a.m. I am awake, and as breakfast arrives I sense a shape on the food tray. I touch a plastic-coated menu and hold it up to my face, breathing in a sweet disinfectant smell. As the image starts to clear, I slowly and deliberately spell out the word 'M-e-n-u'. Looking at Mum in shock, I feel as if I have just conquered Everest.

When I hear Fred's hellos I am curled up in the marsh-mallow chair, but I eagerly wave him over to show him my new trick. As he leans over I gently take the corner of his suit jacket and peer at it closely. It smells masculine and slightly musty. 'This has pinstripes on it,' I state, looking up in his direction.

'You make me very happy,' he simply states; but I can hear the huge grin in his voice. I am vaguely aware of my new boldness, and of how any boundaries seem to have melted away. Last week I couldn't have imagined reaching out and touching the jacket of a consultant neurologist in this familiar way. But, that was last week.

Fred asks me to participate in a junior doctors' tutorial he has arranged. I understand all too well that I am an unusual case study, so of course it stands to reason that I am intriguing in medical terms.

The room Mum wheels me into feels large with muted light coming in from the right. I guess from the shapes that there are a number of junior doctors draped on seats around a table. The room is hot and unnaturally quiet, with a palpable air of tension. *The Big Man* starts to ask the doctors what questions they would ask me as a patient. This is received with yet more silence. Fred quickly brushes aside a couple of stuttered attempts, and – obviously irritated – tells them to make the most of this opportunity.

'You have an exceptionally eloquent patient here who can intelligently answer your questions. So what would you ask her?'

The room is heavy, but I hear someone scribble something, then I guess a student stands up. The young man is around ten feet away from me, so I can't possibly see what he is holding up, let alone what he might have written. I can only tell his sex when he asks me what I can read. I look in Fred's direction and shrug.

'That should give you your answer,' he barks at the figure.

More silence ensues, then I sense a movement by the table. 'Leave that!' Fred shouts. 'You can eat all the biscuits you like once Mrs Potter has left, but please respect that her time is valuable.' At this juncture I have to suppress a small smile, and I can feel Mum shaking with silent laughter behind me.

There are a few more tentative questions, but before I know it I am back in my room.

'Well, he doesn't mince his words, does he?' Mum laughs, obviously impressed. 'I like *The Big Man*. He knows what he's doing.'

'He doesn't frighten me,' I grin.

After the excitement of meeting the students, a sense of hopelessness soon settles over me. I therefore waver a little when the physiotherapist arrives to take me to the hospital gym. Wheeled into a cold, echoing room, she gently assures me that we can stop at any time. When I'm helped down onto all fours, my head instantly starts swimming and I am overcome with dizziness. It takes incredible exertion to lift one leg, then the other, and my whole body shakes with the effort. I have no idea where I am, or if I am facing up or down. Every time we change position we have to wait for my senses to readjust. My body prefers being horizontal these days, so it doesn't yet register diagonal. I am breathless and emotional with the exertion, and the room spins around me. *What has happened to me?* I made it to three Zumba classes last week, but now I can barely lift up one leg.

Helping me to my feet and supporting my full weight, the physiotherapist guides me to my next challenge. Muttering loudly and gripping parallel bars, I now attempt to heave my deadened legs forwards. My unnaturally thick hands struggle to support my weight, and I am unsteady and hesitant. Eventually my baby steps make it all the way to the end, and I stop inches away from a full-length mirror, panting heavily. My breath must be steaming up the glass, but I just stare anyway. An overly jolly voice asks me to turn around, but I can't because I am glued to the spot. Hot tears run unchecked down my cheeks. 'Hello,' I whisper to the faceless ghoul in the mirror.

When I eventually return to my room I am almost instantly wheeled off again for another plasma exchange. I grab my Evian bottle on the way out of the door, clutching it so tightly that I hear the plastic crack. During this session I feel woozy and peculiar, and things I don't understand are going on inside my head. Concerned, Ed leans over me, but as he does I grab his shirt and pull the fabric close to my face. Hovering, teasing me, a faint double stripe hangs in the air. Startled, I look up and take his face in my hands and slowly move it from side to side. If I peer really closely, so closely that our noses almost touch, I can find his eyes – and I smile into them. Our reverie is broken as my fingers start to tingle, and I flex them suspiciously. The numbness has somehow eased without me realising it. It comes back straightaway, however; but it is a change, and not for the worse. These are all changes for the better.

I hear Marissa's friendly call, and her smooth cheek brushes against mine as she hugs me later that afternoon. Her leather jacket is brittle from the cold and I hunger to touch it. In my sanitised, temperature-controlled room, I am unaware of the world beyond my window and that the temperatures are starting to drop. Autumn's bite is taking hold, and I instantly wonder where my children are right now, and if they have their coats on.

Marissa sets to work on painting my nails. We aren't going for the sophisticated burgundy or cherry black I would normally choose. Today we are using orange, electric blue, white and apparently a very loud pink. We have a plan to use sensory stimulation to reignite my brain to help me see colour, to remind my brain what colour is. I don't plan to stay on this misty grey planet for very much longer; I want blues, greens and yellows back in my life again.

I waft my hands around to dry my nails, the wet sensation extremely odd on top of my rubber-gloved fingertips. Suddenly *The Protégé* marches back in unannounced, avoiding the customary knock on the door. As we both look up in surprise he exclaims, 'Ah! Nail painting?' adding as he leans over me, 'Are you making a fashion statement?'

I am mute with shock that he could be so far off the mark. Stumbling over the words we attempt to explain what we're doing, but his footsteps are already moving away from the bed. Blinking in his direction, I start to understand the hospital departmental divides. Our pitiful attempt to stimulate my brain sits more comfortably within the domain of occupational therapy, and this young man with his jangling keys doesn't appear interested in our homespun attempts at kick-starting my sensory system. But, something inside *me* believes that I can affect and impact my own recovery. My curiosity is piqued. Even in my reduced state, I feel that my brain is capable of seeing far more than it actually is right now.

After *The Protégé* has click-clacked his way out of my room, I'm left wondering how any doctor in his right mind could think that a woman plunged into blindness would be remotely concerned about vanity. Marissa and I are shocked, and in his wake attempt some feeble jokes. However, as I settle back onto my pillow it occurs to me that *The Protégé* doesn't see what my family sees. My wild, staring eyes, battle-scarred arms and the Frankenstein tubes sticking out of my neck are not strange or unnatural to him. I am simply a patient, one of the many he sees each day. He cannot compare me with the vibrant, healthy woman who chased her kids in the park only a week ago. He does not know me; he cannot see the drive I have inside me right now to do anything – *anything at all* – to recover. In some small way, it is he who is blind.

Phil, a friend from childhood, arrives to see me at teatime, and it's lovely to hug him and pat his brushed suede jacket, so familiar and comforting. He smells like Phil. I unnerve him a little initially by continually touching his hand, but he soon gets used to it. He's never one for small talk, and we embark on *The Times* crossword. I am not normally a crossword fan, but I am getting more and more drawn to them, and in particular, to using my brain in the logical way they require. It's a remarkable distraction, and somehow normalises me for a short while. However, it's only later, when Phil has gone and Dad picks up the paper, that I discover Phil had misread one of the clues. One cryptic question we'd got stuck on, and that I had been mulling over ever since, was in fact incorrect.

For me, the irony is profound, but I discover a mild annoyance that while Phil has full sight, he is sloppy with it. Of course this is an overreaction, but in my prison bed these things seem to matter. I notice that my own obsessive attention to detail is still present, even without my sight. I am piggybacking the vision of every person that walks into my room, knitting together the patchwork quilt of information that this room provides me with. Yet, it was the blind woman in the bed, with no sight of her own, who eventually uncovered this minuscule and insignificant error.

As I lie in bed I gently flex my feet, having ranged from loud muttering to outright shouting at them today. This one-way dialogue has abated now, much to the relief of my family. I have settled on a more gentle persuasion to get them moving again, and we are still rubbing any manner of strange objects over them. Cotton-wool balls, a string of beads and a nailbrush still litter the end of my bed, brought into my room in the hope that they will spawn hope. The nailbrush in particular evoked a unique agony I could hardy bear. It was like being tickled and scratched at the same time, and I squirmed and yelped as the brush zigzagged over my skin, sending confused neural messages ricocheting around my brain.

Attention to detail, Wednesday, 10 October 2012

I am being quietly observed when, leaning on my stick, I hobble unaided to the bathroom at 4 a.m. Marissa, huddled sleepily in the marshmallow chair, eyes every tentative step I take. It's a small triumph, but my life is made up of tiny triumphs these days.

After breakfast I am quietly scanning my bed to see what new things have appeared overnight. The room is being put together like a jigsaw puzzle – new shapes emerge out of the shadows, enticing me to guess their names. My sight is returning page by page, each one a filmy sheet of acetate with tiny scratches and marks on it. This morning three black smudges on the far wall beckon me over, demanding my attention.

Closer to hand, a Nivea bottle provides a more instant reward. Something so familiar to my morning routine has taken on supernatural qualities. In previous days this dark, heavy shape would have been handed to me, and my fingers guided to the lid. But, this morning I know instantly what it is, and what's more I know there is something else there. It's so faint that I have to hold my breath to register it, but a pale circle rises through the mist. As I hold the bottle two inches from my nose, indistinct outlines start to dance.

This is packaging; this is my job. Examining the fine print on the product shot of a television commercial is part of what I do. Carefully checking we have the right brand, seeking out unwanted highlights and reading every single line of copy is natural to me. I have knelt in front of many post-production monitors to check these small details; but I never thought that I would find myself lying in a hospital bed doing the same thing. A smile twitches at the side of my mouth as I recall the characteristic phrase 'attention to detail' that you'll find on many producers' CVs. Today it has a bitter irony.

It's the last plasma exchange today, the last time I will experience the acerbic commentary of my sharp-tongued technician. I want this magical treatment to go on longer. I want to know if the coiled up animal that lies sleeping on the white T-shirt the technician always wears will today wake up and show me its form.

I have another appointment in Neurophysiology later, and I dread a repeat of the awkward silences of a week ago. Today, I am determined to pass this damn test. As I am again wheeled into the stuffy room, a taut, brisk woman takes the place of my softly spoken friend. The hushed pleasantries I am getting so used to gently buzz around my head, but invisible fingers point silently at me. A bouncy young doctor is overseeing this process today, her bubbling enthusiasm slightly nauseating. I feel like her experiment.

The brisk woman fusses that I should not be allowed to stay in the wheelchair, but is overruled. I can feel her bristling behind me as she glues the sticky electrodes to my head, and am conscious of my unkempt hair becoming even more banshee-like.

Brisk Woman asks me to look at a small spot on a TV screen in front of me, and my heart sinks. 'What TV?' I ask, aware that this is going the same way as last time. The room falls silent again, and a collective yet soundless sigh of dismay breathes in and out. I sense *Brisk Woman*'s body jerk to my right, and I guess she is pointing at the screen. I lean forwards to touch a cold metal square and can only guess this is what she means. The enthusiastic doctor fusses around me, guiding my hand to the small dot. I can feel the technician's disapproval, and her impatience is like a wall of heat behind me.

The other people in the room lower their voices, perhaps thinking that I won't be able to hear them, given that I can't see them. In fact, my hearing is like a fox's at the moment, alert and tuned into any sound, in particular the whispers of clinicians trying to avoid me hearing them. My ears twitch at the words 'waste of time' and *Brisk Woman* finally bursts out.

'She can't see anything!'

That's it.

'No!' I growl from the depths of my wheelchair, turning in her direction. 'I *can* see something.' I start pointing around the room. 'There is a chair over there by the door, and there is a dark shape there on the wall that I think is a coat hook; and you are wearing a skirt.' I breathe furiously as the silence reverberates around me, but I haven't finished yet. The fight in me is rearing up. 'Five days ago I couldn't see *anything*; but today I can see something. Okay, maybe not much, but there is *something there*. So don't tell me I can't see *anything*, okay?'

My tirade diffuses the atmosphere in the room, and I can feel Mum invisibly punch the air from her dark corner, and I taste her delight. Apologies fill the room, contrite and shamed. They forgot I was a person, and now they are sorry. I insist they go ahead anyway, and I pass their infernal test.

'Abnormal readings' is the result, but nonetheless there is some evidence of electrical messages trying vainly to get through. If they'd only asked me, I could have told them that. I am not off the hook yet and my punishment for being mouthy is to be wheeled off for yet another bag of intravenous steroids and vitamins. This is to be swiftly followed by another fMRI, this time with a glow in the dark dye. Maybe when they have lit up my insides like a nuclear Christmas tree they will cross more nasties off the list.

The Protégé arrives later to give us an update. My little gang has a wager running that he is the star pupil here. I can tell from his boyish confidence and smiley voice that he is young and good looking; but he also seems kind, if sometimes impatient. I have forgiven him for his lack of tact about my nail polish, and we have settled into a jovial banter when he now waltzes into my room. He announces that once the fMRI results are in they hope to work out what has happened to me. I have a mini leap of excitement at this brave statement. They might be able to pin a label on me after all, although to date nobody has been any the wiser.

Once *The Protégé* has left Jackie asks how I know that he's attractive (after agreeing with me that he is). She's curious as to why I could say this so confidently, when obviously I couldn't see his face. My answer is simple: 'Because nature would never have made him ugly.'

As I hear my own answer I realise this reflects how we assess those around us in life. High achievers transmit their success through their behaviour. I had calculated from his stride that he was tall, and also noted the reactions of the women in the room when he arrived. It was a combination of these factors, along with more subtle primal cues I couldn't explain, which helped me draw this conclusion. In essence, I didn't need to see him to know he was attractive. Jackie's question left me musing over the complex (and often unthinking) ways in which we humans communicate with each other.

A friend visits me later on, and I smile as I recognise her familiar voice when she kisses my cheek. She is wearing a polka-dot scarf, and I can make out vague stripes on her top. I wonder if Team Family briefed her to wear bold patterns before she visited. She narrates the story of her mother's recent cancer diagnosis, which prompts a thumping to start deep inside my chest. The doctors have eliminated cancerous tumours from my diagnostic list, but bad thoughts drip down inside my mind anyway. We try to cheer each other up when really we're both miserable. Health is such a delicate business.

Richard and Judy, Thursday, 11 October 2012

It was officially a bad night. I haven't rested well, which leaves me worried that it might not be a grain-of-sand day today. But, doing a mental body scan I quickly assess that my fingertips aren't quite as numb this morning. *That's okay, one step at a time*, I tell myself.

Because of this improved sensitivity I can actually leaf through the magazines that I can hear rustling on my bed when I move my feet. Of course, this will just be a habitual act, a normal everyday activity that has opened back up to me this morning. I have no expectation that I will be able to actually read anything. I just want to pick one up and feel the shiny texture of paper between my fingers. I want to do something *normal*.

Idly reaching for the top magazine, I find myself staring dumbstruck at the cover. Pulling it closer still I discover myself nose to nose with a grey, blurry face. The fact that I can identify this muddy, swirling image as a face is amazing! Squinting hard I realise that I know who it is. Incredulously, I stare into the ghostly eyes of Judy Finnegan. I never thought seeing the face of a breakfast-time TV presenter could give me such joy, but this is the outside world creeping back into my life, and Judy Finnegan has just landed on my crazy planet.

Mum and I rifle through the various discarded papers, testing what I can and can't see. Adverts and broadsheet headlines with bold, dark text miraculously drift in front of

my eyes. I proudly and deliberately read out several sensa-
tionalist headlines, the paper held so close to my face that it
muffles my voice. We work out that the higher the contrast
the more detail I have. I don't see any colour yet, but can
distinguish some vague tonal differences. My fingertips are
definitely becoming suppler as the morning progresses. They
now feel less bulbous and cumbersome, and I flex my hands
into fists like a boxer, smiling inside.

I can't wait for the next visitor to arrive so I can tell
them; I think it's fair to say, this is officially a grain-of-sand
day. It doesn't take long for my door to be smartly swung
open, and my nose automatically wrinkles as a wall of
cologne billows across the room. My new visitor tells me
where he has just flown in from, before he tells me his
name. With a sinking feeling inside I can tell this isn't
going to go well.

Cologne Man requests that I follow him down the full
length of the corridor to where there is apparently an eye
chart hanging on the wall. Propping open the door, he sets
off at a brisk march, tap-tapping ahead of me, and I hear his
shoes slide to an impatient halt at the end. The Italian leather
soles tell me he's at least a registrar, perhaps even a consultant.
Hand gripping my stick, I clench my teeth and slowly force
my legs to move in something that resembles a walk. It takes
Mum and I five long minutes to reach the eye chart, during
which time I can hear the impatient swishing of his shoes on
the polished floor. By the time I arrive I am exhausted and
dizzy, but am also determined to read his eye chart. Leaning
on my stick and fumbling in my dressing gown pocket, I try
to find my glasses.

'Don't bother with those,' he snaps. My hand freezes in
mid-air, but I slowly lower it. 'Can you read the middle line?'
he asks me, and I stare at Mum incredulously.

'Have a go at the top,' she whispers, touching my arm.
Squaring on to the wall I fumble again for my glasses and put
them on, and stare at where I think the elusive chart must be.
I stare so hard I can feel my brain ache. There is a suffocating
anxiety twisting itself around my head as my mute

concentration turns to frustration. As I start to take a step closer *Cologne Man* shoots his arm out, 'No, you need to be that distance away!'

'What?' I can hear my mum finally find her voice. 'She was blind a week ago and you expect her to read an eye chart? Does it really matter where she stands?'

Cologne Man relents, but even when I am allowed to touch the thin cardboard, no shapes appear through the mist.

Cologne Man lets out a frustrated breath, and I sense him finger the ID chain around his neck. Turning my head, I seek out the space where I know his eyes must be and whisper fiercely, 'I can see. I saw Judy Finnegan this morning. I just can't see your stupid chart.'

He is obviously taken aback, and I have to admit I am quietly delighted.

'Yes, young man. I think it's about time you lot starting thinking a little more laterally!' Mum blurts out, emboldened by the man's obvious confusion. The moment passes and, somewhat browbeaten, *Cologne Man*, to his credit, apologises and suggests we start over again.

He leaves my room half an hour later scratching his head and smiling at our jokes about how he might clinically record my sight, given that Judy Finnegan's face doesn't feature on the Snellen eye chart.

It *is* a grain-of-sand day.

Click-click-click

My stepmother sits in the corner of my room and I can hear her concentration in the clacking of her needles. My dad settles himself familiarly on the marshmallow chair and I hear a newspaper being unfolded. My stepmother's chatter flits between the imminent arrival of her first grandchild and her next marathon. I find myself zoning out, as while I like being distracted by other people's lives, giving it my full attention is very tiring. I need constant rest. Even a short conversation can leave me almost winded, devoid of any energy and with my body silently shaking in disapproval.

I drift off to my beach, my visitors unaware of my absence. I catch vague snippets of conversation; some miracle cure is being tossed around, an organic remedy available from health-food shops. I hear my dad being instructed to find it, his paper absently rustling in response. As I float here on my beach, I doubt that he will.

I sense that the walking stick, specifically cut to my height, is being subtly caressed as it hangs on the bottom of my bed. I have noticed that people of a certain age can't resist touching it when they spot its pleasantly curved handle. The needles are still click-clacking in the corner, so I smile inwardly at who it is.

I hear Dad say that he's pleased that I don't look like an extra from a horror film anymore. The nurses finally removed the tube sticking out of my neck this morning, much to my relief. I could feel the tube flapping as it came out and was shocked at the length of it – and glad that I hadn't been able to see it go in.

With my Frankenstein appendage gone I am now allowed a shower, this time without water pooling around me on a plastic chair. Joy of joys, I get to hobble alone into my own private space, to wash my hair and to remember what water feels like on skin. The utterly wonderful feeling of hot water pouring over my head is bliss, and just being clean again is humanising. Who knew shower gel could be so fabulously smelly and evocative? The shower room fills with steamy pomegranate and sighs.

The jolly physiotherapist is back again with an ankle strap to stop my right foot from collapsing when I walk on it. I show her the tiny wiggle I can now manage in my right toes, and she claps her hands in delight. Fingering the cumbersome ankle strap before it is fastened around my foot, I privately decide not to use this for long. Obediently I practise walking, even managing to manoeuvre up and down the corridor, impressing everyone with my explorations. My ears prick up when I hear them talking inside the room.

This is good progress. It's much more than we had hoped for. Standing in the dark corridor, this is the furthest I have gone on my own two feet and I am amazed at how mobile I feel. I don't want to stop.

As I return, grinning, the first mention of me going home is mooted and my mouth falls open, then I burst into uncontrollable tears.

I don't understand why they said it. I couldn't go home like this. If I go home in this blind old woman's body, I will stay like this forever. *Don't they understand?* I am trapped inside a macabre fairy tale, but I don't want to stay the horrid old witch forever. I need the magical potion and the spell to be broken. Recovery takes place in *hospital*, not at home. The magic transformation needs to happen here. I'm only going home when I am the right size again.

Listening to the murmured platitudes humming around my head, it starts to dawn on me that this may not be possible; but through my wall of tears I just keep repeating *no, no, no*. I want the happy ending.

Needless to say, the latter part of the day isn't quite so positive. I have been bombarded with visitors who Ed had enthusiastically booked in. Two friends are already here, but pinging texts alert us to two more work colleagues wandering aimlessly around the hospital. I am double booked.

This is overload. No one seems to understand that while I am a very social person normally, I am not *myself* at the moment. I can't *see* anyone, so therefore I can't see anyone. *Why can't they get that?* Even a short conversation with friends I know leaves me a trembling wreck, desperate to dive back into my watery haven.

With my eyes now tightly closed I wave the new visitors away, swatting at their words, their pity. My current guests both happen to be nurses; and with an ingrained knowledge they see what I am. They see a patient. This is what I have become. I am no longer Vanessa, so they slide into familiar roles. Ed is diplomatically ushered out, and the work colleagues politely but firmly sent home. A warm hand slides silently into mine, and I spend the evening in a giant hug as my tears flow silently and unchecked.

Cup of tea, Friday, 12 October 2012

I couldn't extract a train-crash pill from the nurses last night and, yawning, I lament the resulting bad night.

The Protégé struts in waving the second fMRI results, and I hear him pause by my bedside and scratch his chin. His voice is tinged with disappointment as he announces that the scans were normal, admitting that they were expecting lesions, or something to be visible. He rambles on that they have requested some antibody tests, which will take several weeks to process. These may or may not confirm an NMO diagnosis.

I am instantly confused. *Surely the fMRI being clear is a good thing?* I cannot understand why all my negative results are, well, negative. When I pose this question he chips in quickly, 'Yes, yes,' but his reply is completely unconvincing.

He manages to shrug off his evident frustration, but I sigh as I realise I am a number to him. I am the diagnosis he may not be able to make, the prize he may not win. But I am also relieved. We are ticking things off the big, scary list, and I want every test to be clear, even if that means they don't know what happened to me. It would appear that for today at least, I am the mystery patient once more. I vaguely hear some new terrible illnesses mentioned and I mentally switch off. Somehow cat-scratch disease lingers with me, but I put it out of my mind, as this nasty has also been ticked off.

The world is full of people making inappropriate comments in the workplace, and as I have discovered, hospitals don't have immunity. I am hugely frustrated when these misplaced comments come my way. Lying down, I don't feel equipped to hurl back the required cutting retorts. My family are some of the worst culprits, making jokes and witty observations in an attempt to lighten the situation. *If we can laugh, then it's all okay, isn't it?* If Ness can see the funny side then it can't be that serious. I play along, mostly, beginning to understand that they need this charade more than I do.

The going home chant is dominating conversations, but there are a few more tests that I have to pass before I am allowed out of this place. I am being heaved into the wheelchair and guided to what is described as a makeshift kitchen that appears to have been constructed at one side of a room.

I'm told it is necessary for me to show them I can make a cup of tea. I feel hysterical giggles start to bubble up, at how incredibly British this all is. It is so typical of this country to make our patients make a cup of tea. Do Italian patients get to make a cappuccino?

My humour is soon dispersed, as I understand quite how hard this task will prove to be. Hoisted out of my wheelchair and holding onto the worn formica, I am instructed to identify where the cups, kettle and teabags are. Sliding my feet up and down the floor, my toes touching the plinth, I try to get my bearings. I can see nothing in front of me; the grey, swirling mist has swallowed up everything. Already breathing heavily, I tell myself that as long as I hold on, I will be okay. Fumbling along, my rubber fingers feeling the way, I touch a handle and slowly drag a ceramic mug towards me. Cylindrical shapes loom somewhere out in front, and I lunge in their direction. I hear the sound of tins rolling as I sway back, dizziness engulfing me. I feel as if I am on a ship in a stormy sea, but warm fingers touch my elbow and reassuring words brush my ear. My cappuccino joke feels a long time ago now.

It was possibly one of the hardest things I have ever done, physically and emotionally. However, I eventually produced a cup of milky, unappealing tea – and began to realise that there were many more hurdles I had yet to leap.

A sprinkle of heaven, Saturday, 13 October 2012

Dad was on duty last night, but he was distracted making travel arrangements and sat muttering at his laptop for much of the time. For some reason he put on the light in the middle of the night, and the room lit up in a way it had not done before. I was dazzled, yet it was thrilling to experience such a big shift in light. I couldn't register any additional detail or form, but it did make me realise just how much light I have been missing.

In the morning, sensing Dad's dipped head, I decide to test my ability to shower by myself once again. The tiles are wet so I take tentative steps inside. Feeling my way along the walls I attempt to find the shower controls. The ceramic feels horrible under my thick, clumsy feet, stiff after a night's sleep. I don't have enough grip or agility yet to trust my step, so I am terrified of slipping. Without assistance this morning undressing is awkward, so I devise a system that will mean everything will get soaked, but I don't care. When I am balanced I turn on the shower and yet again it is as though heaven has burst open from above. The water is not just warm and comforting, but also healing.

As I open my eyes I am suddenly aware of dazzling beads of light scattering as I move. Astonished, I see the tiny water rivulets streaming down from the showerhead. Choking and spluttering I shout out to Dad through the closed door, 'I can see the water jets!'

It is a miracle. I can't see my body or my own face, but I can see these thin, glittering streams of water.

Feeling buoyed and wanting to celebrate, Dad and I head off to the hospital cafe. These expeditions are frightening and disorientating, and I can't help but fire a constant barrage of questions all the way there. Even seated at a cold metal table, I ask Dad to describe the full scope of the room. My voice seems overly loud and unnatural, and I feel alien in this new place. I point to two dark circles floating on the horizon, and Dad tells me they are the holes of the recycling bins. Even this limited stimulus is exhausting, and after a few minutes my body starts shaking. The small triumph of popping off the plastic lid from my hot chocolate myself (proving my fingers are more flexible today) is lost amid the noise and confusion of this place. I know I have to make myself stay calm. I am all too aware of how vulnerable I am, and I crave the safety of my beach.

Back in bed I stare at the iridescent hospital wall, erratic and headache inducing with its shifting shapes and tones, and I am suddenly filled with a longing. It is a longing so deep and severe that it winds itself around my chest, tightening and constricting like an invisible belt. All of my senses are being bombarded; my whole body is on red alert and a silent voice screams inside my head. I am forced to look through this quivering celluloid window, which every day hints at more, but never clears for one moment. The flickering is exhausting and makes me agitated. I long for my old life back, for what everyone who walks into my room has without question – *perfect sight*. It is a pining, an ache, and it is audible in my breath.

I miss colour. I miss everything, but I *really* miss colour. My eyes have always been lenses to me in the truest sense of the word. I saw life as a series of beautiful microcosms. Photography wasn't just a subject I studied at univerisity; it was how I saw the world. It was how I perceived and took meaning from the subtleties, the flashes of movement and the unrepeatable glimmers of life around me. Composition, light, tiny facets of detail; it was the dirt in the cracks of the pavement that intrigued me, that drew me down onto my knees to really look at them.

Do you know how many shades of red there are? How far magenta has to go before it slides into pink, or how far it has to retreat back into a safer, more conservative maroon? Colour has personality, meaning. I know there are others who see colour in the way I do, but they are not here now. We know colour talks, and sometimes it even shouts.

I can say all of this because today I *made* myself see blue. It didn't happen spontaneously. I had to study the muted, wishy-washy images and find it – to hunt it out in between the monochromatic shapes that mutate in front of me. I had to wait for blue to whisper my name. It wasn't the same blue everyone else sees, of course; it was just a hint, a bluish tint to the muddy greys all around me. If I hadn't concentrated quite so hard, I could easily have missed it. It happened when Jackie and I ventured outside to catch the last of the afternoon sun.

She pushed my wheelchair, as my name placard flapped like a trapped bird behind my back.

We didn't just keep to the synthetically lit corridors today, but instead we emerged into the real outside, the place with sunlight and busy people carrying babies and briefcases. As soon as the automatic door shushed open, I was hit with a sword of white fire and bolts of luminous jets seared my face, my head and my brain. The intensity was breathtaking. I sat silently blinking my unnatural owl eyes with their overly large pupils, soaking up this glimmering white world.

Ghosts and wispy spirits float past me and I rubberneck to watch them dissolve into the air, fascinated by their translucence. I sense us pass close to a large vehicle and instinctually put my arm out to stop. Leaning precariously out of the wheelchair, I flatten my hands on the dirty metal, aware there is something there. Slowly, letters appear one by one, 'a–m–b–u–l–a–n–c–e'. I know I must look like a crazy woman reading out each letter; but I don't care, this is incredible.

Balloons are tied to a bench next to the coffee shop. '*Stop!*' I shriek as we bump past them. Pulling down one of the ethereal shapes from its floating position, I study it with stern fascination. I was sure it had made me stop. '*You* are a colour,' I tell it firmly, 'I know you are.' Aware of the stares and glances from the passing ghosts, I am unabashed and rebelliously tuck it under my arm. This is my balloon now because it has just murmured my name. As I glance down at my hands I can feel my nail varnish starting to peel, but there are different shades and tones there. They aren't colours yet, but they are quietly mumbling to me.

I'm sure that from a distance Jackie and I looked like a scene from *Little Britain*; certainly, my odd behaviour and random shouts drew some attention. Even at the best of times wheelchairs are a demeaning mode of transport. They leave you at the mercy of your carer, not just physically, but mentally. They also, ironically, don't fit inside the hospital shop we were headed to. I am therefore parked outside and left to my own devices. Looking around I force my unseeing eyes to orientate myself, but walls rear up and a shadow jabs at my face like an erratic boxer. I realise I can stare directly into the sun without squinting, which is a strange sensation. On a bright, sunny day like this I would normally be huddled behind my dark sunglasses, squinting even then. The sun's warmth on my bare arms is like a cure-all. I feel uplifted and free from the confining walls of my room. My horizons have just got much bigger, even though I can't quite see them yet.

On the way back inside I notice my stone-clad feet starting to break up. They are crumbling inside my thick woollen socks. I wriggle my toes a bit and smile – perhaps some of the shouting has been worthwhile! For the first time I realise I want to go home. I can manage.

Whenever the overwhelming exhaustion comes over me, I know I must lie still. I don't sleep, but my body is still in the truest sense of the word. I lie at peace in my quiet inner place, but this time an electric storm erupts inside my head. This storm is not about exhaustion, though, and as I lie motionless lights flash and explode inside my mind like fireworks. I have no experience to compare this to, and fear starts to tickle my neck. I am getting used to strange bodily sensations, but this is in a league of its own. Fumbling around, I drag Ed's sweatshirt over my head to block out the light, but the flashing continues. My head is floodlit from the inside, and I have a grinding headache. Frightened, I discover that I cannot physically open my eyes. The noise inside my head is so deafening that I wonder if Ed can hear me whimper. I wave frantically for him to come. I need to feel his warm strength next to me. I need to feel his beating heart to keep me rooted here – to stop my mind from exploding into tiny pieces. Ed hasn't lain down next to me at any time up until now, but suddenly I crave his closeness, the familiar smell of him. He doesn't know what is happening, either, so as he wraps the sweatshirt tightly around my head we both just wait it out.

About two hours later the lights suddenly stop and I sit up and pull the sweatshirt off my head. It's over.

'What the hell was that?' I ask, but Ed has no idea either. As I look around all of my sight has been washed away. The tiny fragments of detail I had this morning have evaporated, and my room has been replaced with an old sepia print faded with age. I know I have to wait for this to clear, so I try not to panic. I instinctively know that it will pass. Clutching Ed's warm hand I breathe my way back to my beach – to where there is colour and warmth. I flood my mind with vibrant hues once more, and paint my world back in again.

Children, Sunday, 14 October 2012

Our son and daughter are here.

It's a bright, sunny day again, and I am excited but nervous to finally greet my children. The only thing I focus on is identifying their white-blond hair. I don't expect to see anything else.

My daughter shyly approaches my chair, but soon climbs up and snakes her legs around my waist, gripping me tightly. My son runs up and bashes me affectionately. He resembles a puppy more than a boy, but his untamed energy is infectious. Both children circle me suspiciously, gently touching the bruises on my arm, the plasters on my neck, and questions tumble out at a rate I can't keep up with. They are curious as to why I am in a wheelchair, why I am here. They know Mummy doesn't look right, but they are pleased to see me nonetheless – besides, I sound and smell like their mummy, so that's okay for them.

My daughter is particularly concerned about the huge plasters on my neck. She knows all about plasters, but these ones look scary and too grown up to her. I reassure her, saying that they don't hurt, but she is not convinced. Plasters on adults are an unfamiliar sight in her world. The ones she knows about have Peppa Pig designs on them and only belong on children's knees.

The meeting goes well even though a panic swarms over me intermittently as the children dart in and out of my visual range, disappearing like spirits into an invisible infinity. I grit my teeth and constantly ask Ed where they are, but he is his usual nonchalant self. I realise that as a family unit, I am the one who does *the watching*; and even in my reduced capacity, that job is still mine.

CHAPTER FOUR

D-Day

Memory 3
A Royal Wedding, July 1986

It's the summer of 1986; the school holidays offer endless lazy days and the promise of adventures and freedom. My friends and I mill around Glen Road, a quiet residential lane in West Yorkshire. We spend the days skateboarding, biking and generally hanging out, our shoulders turning a little pinker every day. Life is just better in the summer; the days smell hot and arid, of melting tarmac. There's a different quality to the air, a low-level thrill that isn't present at any other time of the year. Every day leading up to the end of term had been torturous; I yearned for school to just finish, for the holidays to explode into life.

I awake to birds singing a riotous chorus outside my window, knowing that school is finally over. As I drag back my faded curtains the garden bombards me with loud primary colours. The BBC has been reporting daily on the upcoming royal wedding of Sarah Ferguson and Prince Andrew. For teens and adults alike it has become the sole topic of conversation. My excitement surrounding the wedding has been building, and I'm itching for an event, a party – for just something to be going on. Having just turned fourteen I am thoroughly caught up in the constant speculation about the Emanuels' wedding dress, and with my indistinct mousey hair that is neither curly nor straight, and my slightly solid frame, I welcome the romance and excitement.

It was the most natural thing in the world to throw a street party, and as far as I was concerned I should be the one to organise it. This could have been a large undertaking for a teenager, but organising events wasn't something new to me. Christmas carol concerts had taken place over the years, with my stilted, childlike attempts to play the piano accompanying my forever-loyal gang of friends as they trilled 'Silent Night' to quietly amused parents and neighbours. Detail was important, so my friends were forced to wear woolly hats for a truly genuine carol-singing look, while I was unsympathetic to the surreptitious scratching and mutterings that ensued.

It didn't take long to organise this party. I settled on my grandparents' driveway and garage (in case of rain) as the location. Food was easy; I wrote a list of what kind of sandwiches we'd need, what drinks, crisps, salads and cakes, and of course being a child of the Eighties, I did not forget the Martini Bianco and lemonade. There wasn't a detail I didn't think of. Knocking on my neighbours' doors, I handed out a list of what they should bring. To my credit, I was inclusive. I knocked on everyone's doors, even the ones who disliked us noisy children, and the people whose houses smelled funny. The first my mum knew of it was when she received several bemused phone calls following my canvassing expeditions. Nevertheless, she sighed and told the neighbours to bring whatever they had been asked to bring.

The party was a complete success. Everyone dressed up in red, white and blue, and the food was a fanciful feast of wobbly jellies, pineapple and cheese sticks, and perfectly tiny sandwiches. I, along with several of the local boys, commandeered a squashy bottle of cheap cider and got gently sloshed in one of the deserted gardens. It was an event to remember, and everyone was happy. I didn't know it then, but that was my first proper production.

Monday, 15 October 2012

In order to be released from hospital tomorrow, two things have to happen today. The first is that *The Big Man* has to see me again, and the ponytailed physiotherapist has to watch me walk down some steps. You would think that these two things would be simple, but they prove not to be.

Pacing around my room (as much as I am able to pace), I practise what it might be like to walk down a step. I try lifting my foot up off the polished floor, but this is almost impossible as my muscles just do not seem to understand what my brain wants them to do. My gait is uneven and resembles that of an elderly arthritic person, and annoyingly my right foot still collapses when I get tired. I worry that I won't pass the stairs test and might be made to stay here. How can it be that something so normal, so pedestrian, is causing me to sweat with anxiety? I ask myself for the hundredth time – *how did this happen to me?*

Eventually I find myself teetering at the top of an alarmingly steep and long flight of hospital stairs. They are so long that I can't see the end point, and I wonder how on earth I am expected to make it all the way down. Instructed to use the stick as a support, I step off and the memory of Indiana Jones taking the leap of faith immediately flashes through my mind. It's a weak joke, but stepping off, my foot hovering in mid-air, I can only hope that the stick lands on the step below. I don't want to drop into the abyss beneath. The relief when my foot makes contact with the hard step makes me whoop with joy. I know if I can do that first step, I can do the rest. I have all the time in the world.

Fred finally appears as the sun is dipping down over the balcony, and I try to follow him as his silhouette paces around the room. He doesn't say much but he does give Ed and I a label to hold onto, which turns out to be NMO. Except I'm by no means a classic case and Fred brushes aside any further talk of the condition. NMO is all I get. However, gripping the arms of the marshmallow chair, I hear another phrase I know I will repeat many times.

'You do not have permanent damage, and you have a high risk of a full recovery.'

I find the use of the word risk strange – I didn't think getting better was a risk, but perhaps that is the best way to describe it.

Getting out, Tuesday, 16 October 2012

This seems seriously like getting out of prison, although of course I'm not an authority on that. The waiting has agitated us all as it feels endless.

The hospital pharmacist comes to see me to check what medication I am taking home. I ask for more train-crash pills,

which she baulks at. She primly informs me that I am not
allowed to take these pills home as sleeping aids. She suggests
that if I wake up at night, I should get up and read a book. I
don't deign to reply to this tactless suggestion, but I do suffer
delayed shock later at the realisation that she had obviously
not read my notes. You would have thought she might have
noticed my incredibly dilated pupils, or that I couldn't look
her directly in the eye, but perhaps no one looks a hospital
pharmacist in the eye.

Eventually we are given the all clear, and bags are stacked
high on the wheelchair. Ed squeezes my hand, then efficiently
manoeuvres me out of the door without a backwards glance.
We are both glad to leave room three. Once again I feel
vulnerable as I am wheeled into the open. I know people look
at me – they see my huge eyes. I look blind.

To my utter horror, Ed leaves me sitting outside in a leafy
quadrant to go and move the car closer. I perch precariously
on a slatted bench waiting for him. I can make out dim lines
as he lowers me down, and can feel the reassuring touch of
wood. I wait nervously, aware that strangers are moving
nearby, frightened that they may approach me, or worse still
try to talk to me. I can make out very little, but a hissing
sound suggests saplings gently swaying in the light breeze
close by. I think about how I must look to everyone around
me – a sick young woman covered in bruises.

Ed and I inch down some stone steps leading to his car.
This is the largest number of steps I have managed so far,
and it is a mammoth task requiring much coordination.
Not surprisingly, we grin triumphantly when we get to the
bottom, but when I see the outline of Ed's car it is painfully
familiar. Slowly levering myself into the seat, my limbs are
awkward and wrong.

There is a thick fog covering the front windscreen, so
I give up trying to see anything through it. However, as
terraced houses and trees flick by, I glimpse tiny fragments
out of the side window. Edges of buildings are lit up by
bright sunlight, an odd lamp post appears, a passing car; even
a drainpipe is detectable as we whizz by. The sun allows an

added layer of sight – there are no details, but there is a delicate world out there, as sheer as a spider's web.

Ed tells me when we hit Crown Dale and suddenly I know we will turn right in a moment, then left. I know these roads, yet I can't see them. The rocking of the car and the surge of the accelerator is so normal, yet even though my body knows this journey, my eyes do not.

This is not the world I left behind two weeks ago. I could drive around those roads for myself then. This is a substitute place, a temporary image. Someone has removed the substance of my life, and all that is left is this wafer-thin membrane.

I close my eyes, feeling that I have never seen so much, and yet so little, in all my life.

An episode

In all, I suffered what the medical fraternity call 'an episode'. It's a perfect description, actually, as it turns out that the episode, which turned out to basically be a very rare neurological illness, lasted a considerable length of time. It became an episode in my life, an ongoing, repeating drama. Overnight, my life dissolved into a fantasy film.

I banned the use of the 'b' word for a long time. The doctors explained that I had suffered profound bilateral sight loss. Indeed, it was what it said on the tin – complete loss of sight in both eyes.

Zip.
Nada.
Zilch.

Losing all feeling and function in my hands and feet only compounded matters. Within a matter of days, two of my major senses had been knocked out cold. My ability to not only receive stimuli, but also interpret and understand the world around me, was deeply impaired. I felt a little like my son's toy box as he tipped it upside down on his head, the contents strewn casually across the floor. I was nothing more than broken doll limbs and the plastic tat that blocks up the vacuum cleaner. There's no way you can prepare for your whole presence to be suddenly sucked into a black hole. Illness hit me like a high-speed train – fast and absolute.

Devic and the Devil

It's worth getting a few facts straight here, as far as that is possible. NMO (neuromyelitis optica), previously known, rather more alluringly, as Devic's disease, is a very rare autoimmune neurological condition characterised by attacks (relapses) of inflammation in the optic nerve or the spinal cord, which can spread to other parts of the central nervous system and brain.

Dr Eugène Devic, a French neurologist working in the nineteenth century, originally identified this unusual neurological illness. It was his work investigating this complex disease that helped it acquire the name; however, today it is most often referred to simply as NMO.

NMO is a serious condition affecting the nerves of the brain, and it can leave those affected with visual impairment or blindness, and even paralysis. I didn't have classic NMO, but my symptoms were similar (if rather severe) as they affected both my optic nerves and my spinal cord simultaneously. Therefore I was considered to be suffering antibody-negative neuromyelitis optica spectrum disorder (NMOSD). My illness was difficult to identify as I never quite ticked the NMO diagnostic boxes, so I sit precariously somewhere on the NMO spectrum.

Most people with NMO (approximately 75 per cent) test positive for proteins within their blood called anti-aquaporin 4 antibodies (AQP4), and these are the cause of the disease. If you have an autoimmune disease, your own immune system turns aggressor and attacks your body tissues. Normally, your immune system protects your body against infections caused by bacteria, viruses and other pathogens. It cleverly recognises when something foreign enters your body, and can usually get rid of it before it causes you any harm. However, if you have an autoimmune disease your immune system can make mistakes, and it sometimes gets things wrong. In these cases your immune cells start to attack your own normal body cells, causing inflammation. In my case, my customary helpful antibodies overnight became my own assailants. As I don't wear a white lab coat for a living, it took me some time to get my head around this disturbing concept.

Another problem with NMO is that while it can be monophasic, it is more often than not recurrent. This means that you can experience a one-off attack, which can affect both the spinal column and optic nerves simultaneously, then remain in permanent remission, or it can recur again and again throughout your life. Fortunately, because the causative antibody could not be detected in my case, my condition is less likely to relapse. NMO is most common in people of Asian and African descent (of which I am neither), and it can occur in children. It is also more common in women than in men.*

The brain is a highly complex organ, influencing the majority of our bodily functions. The connections between the brain and body are like the lead from your kettle, simply millions of insulated electrical wires. During an autoimmune attack on the central nervous system, the insulation around the wire (called the myelin) or the wire itself (the axon) can be damaged. When an inflammatory attack damages the insulation, the damage is referred to as demyelination. When the myelin or axon of a neuron is damaged, it is unable to send signals. Of course, this all depends on what area has been affected; if the wire that carries visual information from the eye to the brain (optic nerve) develops demyelination, signals are not transmitted to the brain efficiently, resulting in a person having blurred, reduced or sometimes lost vision. If the demyelination occurs in the wires sending motor signals to a person's legs, the person experiences muscle weakness and difficulty in walking. Even though autoimmune diseases

---

* NMO affects four or more women for every man.

are varied and affect many people, the exact reasons why our bodies attack our own cells are not yet fully understood.

Finally, while it is difficult to estimate accurately the number of NMO or NMOSD sufferers in the UK due to limited studies, research undertaken in continental Europe estimates that there is about one case of NMO for every 100,000 people, implying that around 700 people are potentially affected in the UK.*

While these figures are for only definite NMO and NMOSD, they do give an indication of the rarity of this condition. Considering the rate of recurrence, out of those 700 people affected by NMO in the UK, 672 most probably experienced recurring attacks, while 28 probably just had the one (monophasic) episode.† To illustrate this rarity of the disease, it's not too far off the mark to suggest that 0.00004 per cent of the UK population has monophasic NMOSD at any one given time; and I just happened to be one of these people.

Roundabouts

However big an event may be, I don't believe that we are defined by one event alone – even if it is a devastating illness such as the one I experienced. There is a constant accumulation, a myriad of experiences that build up like sediment and over time form the person we are. So, while my episode has without question had significant effects on my own personal metamorphosis, it has not been the only influence. In fact, it was its devastating effects on my body that drew out inner resources which until that time I didn't know I possessed. Perhaps the kernels of insight were already there – the faint signals I sometimes caught in the periphery of my awareness,

---

* NMO is a very rare condition, and there are few population studies. In continental Europe, it is estimated that there are 0.5–4.5 cases of NMO for every 100,000 people, potentially affecting approximately 350–2,880 people in the UK.

† The UK population was approximately 64 million in 2013.

a fleeting moment of calm in among the maelstrom of work. However, I suspect I was just too busy to acknowledge them fully. I spent most of my time leaning back, hair flapping in the wind, riding my forever spinning roundabout. I never took the time to understand these flickers of intuition, for I was concentrating hard on the rhythmic beat of my foot as it drove my roundabout faster. I never stopped to consider what was happening on the inside, for I could only see the world as it spun past me on the outside.

It's only when I was forced off my roundabout that I could see how my life's experiences had built up, and what they had become, and in turn how they might save me. Time for once was what I now had in abundance.

CHAPTER FIVE

Home

Memory 4
Auction, July 2007

My palms are embarrassingly sweaty, and I'm not sure where to keep wiping them without seriously drawing attention to myself. I can feel heat radiating off Ed, too, but then he's always hot. The room is one of those tired banquet rooms you find in three-star hotels; gold spray-painted chairs and blue chintz sigh wearily around us. The chairs have been lined up in front of a raised platform, and around half of them are already full of expectant bodies. There is a palpable air of suspense in the room. I cannot drag my eyes away from the auctioneer as he preens his greying hair and sets up his stage, aware that all eyes are on him. Our lot is around a third of the way through the list. I clutch the paperwork on my lap, dog-eared with my constant referencing. A lull at work has meant that I know everything there is to know about this house. We found it by chance while looking at another house three weeks ago. Our lot has a tenant living in it without an eviction date, and we have only seen the property from the outside so we have no idea of its condition. We have no survey, and no other documentation to support buying it. As the current tenant is a mother with three children she has rights, and as such, she has refused anyone access to view the property. There are additional fees and all manner of questionable elements to this purchase, but I have done my homework and assessed the risks; so we are here.

I find myself yet again replaying the events of the previous evening. Driving over to the lot after work, Ed and I walked up to the front door and, asking me to pause five paces behind him, Ed knocked. At six months pregnant I was visibly rotund, so I felt awkward and on show. Fidgeting behind I didn't hear the full conversation, but I could tell from Ed's demeanour and the

*woman's body language that his calming way with people was
having an effect.*

*As we walked slowly back to the car I bombarded him
with questions, but after an infuriating silence he simply said,
'I asked her whether if we bought the house at auction she would
move out.'*

'That's it?'

'Yup.'

*'You didn't ask to look inside? Or what's happening with
the eviction order?'*

*'Nope. I saw no point. She saw you and asked how long
before you were due. So I knew we would be okay.'*

*A nudge from Ed jerks me back to the present and he
whispers, 'Are we doing this?'*

*He gets his answer when my hand shoots straight up at the
first bid.*

*Walking into a film-edit suite an hour later I am distracted,
and bat away the lively remarks about my tardy arrival. My
daughter is kicking her tiny feet deep inside me, and I can feel
her excitement, too. Pulling on my professional face I pick up the
script, and a strange tranquillity comes over me.*

*'Okay, guys. Sorry I was late. I just bought at house at
auction. So, let's see where this edit is at.'*

The Hair Flicker

It's a cold afternoon when Dad welcomes me home with a
hug. I lean heavily on my stick, wandering around the rooms,
touching surfaces, trying to connect with the place I live in.
It feels wonderful yet alarming to be here. The house looks
like I expect it to, but at the same time it is a supernatural
replica of the place I left. The bits of furniture that I can make
out are where they should be, but I can tell people have been
here, inhabiting my home in their own ways. There is so
much that I can't see, and as I strain to understand the shapes
that loom up in front of me, my memory fills in the gaps. I
only know where things are, because I remember where they
should be.

A massive painted canvas of a woman eating a cream cake that I bought more than ten years ago catches my attention. I know it has a vivid blue background, and peering closely I can somehow imagine colour being there. Something inside me is screaming *blue*. My brain is performing neural acrobatics, but all it can transmit is a soulless grey, and I feel a battle waging between what I know and what I see.

I consider describing my strange visual world, but I have no words. Instead I wander around touching objects, muttering the same thing over and over again. All I seem to see are lines. They dominate my patter as they freakishly dominate my view. Wobbly lines traverse the room, telling me where our fitted wardrobes are, hinting at windows, tables and door frames. I am even drawn bee-like to lines on people's clothing, as these vibrating shapes demarcate this strange, misty world. In particular, verticals jump out at me – actually they do more than just that; they assault me head on. I am hypnotised. My visual landscape is a matrix, a criss-cross of contrasting, angular shapes. I know this rigid framework defines everything that I see, and it demands my attention. Yet this visual structure is invisibly blended for everyone else; their brains know *not* to notice what I see. I am forced to live with this raw, unfinished view, one being rebuilt – a visual no man's land.

The strangeness doesn't end there, either. My feet are still freezing cold, and recently an invisible pest has been sticking tiny needles in them. It reminds me of when I used a Tens machine some years ago, except that this time my sadistic friend has turned the dial right up. I am grateful that my brain manages to unscramble this twisted message before too long, as I don't think I could have stood the painful twitching much longer.

Words have new meaning now; they are loaded guns firing intermittently around me. I am home, in a place I know, but so much is alien to me. Even the platitudes that fall out of people's mouths feel so much louder now that I cannot see. I wince as every throwaway word hits me; *sight, vision, blind*. Metaphors litter our language without us even being aware of them, but they have sinister undertones. I must learn to duck their erratic trajectories, for even though they can cut through the fog that enfolds me, I must bat away their significance. They are only words.

Within days an unfamiliar person has come to visit. She has plonked herself on the sofa too far away for me to see any facial detail, so I have little to go on. Her voice was upbeat and full of annoying chatter as she breezed in with Mum. I know she has long, dark hair because I can see it being flicked around her head as she talks. I get odd glimpses of her profile as she shifts in her seat, and while I don't like the reason why she's here, I have to admit that she has a reassuring aura to her.

She is telling us about full-time care, and all I can hear is the slapstick ring of a cash register. I know I need help, but I don't want people I don't know doing it. My concrete toes dig into the rug and I refuse to feel pressurised into agreeing to strangers having rights to my home, my children – a key to my front door. It's all very well that everyone just wants a solution to the problem; but what they're really saying is that *I* am the problem. I don't understand the need to involve others. *Am I becoming such a burden?* Perhaps this is what it feels like to have our children gently coax us into an old people's home when we're too old to think for ourselves. Fast-forwarding several decades, I decide to be as feisty about that, too, when the time comes. I'm suddenly reminded of my own grandmother's reluctance to be cajoled into a home – and for the first time I understand what she felt.

My life is being taken over in other ways, too. There is a fresh, sharpened fear slowly paralysing my body in an entirely new and horrible way. I know it's a foreign, almost parasitic emotion taking over my senses, yet right now it feels more real than anything. *It*, this mythical fear, has an identity all of its own. It is seeping into my mind, rearing up suddenly, then slithering away like a predatory reptile. It occupies my thoughts even more than the sight loss, breathing out foul air that chokes me.

I know this is happening because I am a fairly rational person, and I'm also deeply practical. I understand a little about mental unease, and know how anxiety can dominate you if you let it. This isn't the brief 'I've lost my keys' kind of anxiety, either. This is another beast entirely. Anxiety attacks you, yet it is an invisible assailant.

I can't fully explain this to anyone; there is no way to describe how incapacitating these feelings are. Even walking around or lying down has my stomach in knots and my heart pounding. Catastrophe engulfs me. I hate myself for being so reliant on my family, for needing one of them around me at every minute of the day. I have to *do* something.

These feelings are more intense than they were in the hospital. My beach hideaway gave me a safe place to escape the pain of reality, but I find myself riffling through my mental kit bag again, pulling out an assortment of tricks and tools. I try and recall breathing techniques, visualisations, meditation, but I fear these are pathetic little sticks being thrown at a giant bogeyman. Shaking my head I make myself remember, forcing myself to reconnect to the part of my psyche that I know can help me. Fear won't paralyse me; I've found inner resources before that helped me fight back. A surge of determination floods through me as I steel myself for the battle ahead. I know this will take practice and repetition, but I have conquered fear before. At least by trying, I know I am doing *something*. Okay, so they gave *The Hair Flicker* a key to my front door today, but they can't have a key to my mind. I know the only person who can save me, is me.

The door bangs shut as *The Hair Flicker* leaves, and shortly after Mum goes out to collect the children and an eerie silence settles over the house. The doorbell rings moments later and my heart jumps into my mouth. *How can I possibly answer the door myself?* I panic for minutes before hobbling uncertainly and opening the door a crack. A thin voice informs me that it is my doctor making a home call.

Her whole body is taut and uncomfortable as she perches on the edge of the sofa, and her voice barely disguises a sympathetic horror. She freely admits she knows next to nothing about my condition, and waves of discomfort roll over me. The fact that I have my doctor visiting me at home tells me how serious all of this is. Happy to do something, she scribbles out prescriptions for more, new train-crash pills. She mirrors the same platitudes as every other professional I have spoken to.

'It's a waiting game.'

If only they knew how long waiting takes, and how much energy it uses up. Everyone tells me to wait, but no one ever dares tell me for how long.

As the door shuts I lurch with the knowledge that I am alone again. I can't believe that I feel like this, this helpless shell of a person. I haven't been alone for more than a few minutes in twenty-two days straight. That's 528 hours with someone by my side caring for me. The worst of it is that right now, I can't imagine it being any other way.

16:53

As my brain cannot fully process the light messages it is receiving, there are times when my eyes feel pretty useless. Staring at an inanimate object, knowing that it is there yet willing it to appear, I feel as though I am simply mimicking the act of seeing. My eyes don't feel connected to my brain; there are wires that have come loose. I know our ability to see involves a multipart process and I am beginning to acknowledge the complexities of this.

Staring hard right now, I can just about decipher a glowing shape centimetres from my face. I have to peer through the clouds that hover inside in my head, shifting to get just the right angle. I am so close to the clock that my nose keeps bumping the glass. Three weeks ago these oversized numbers would have been visible from the other side of the room, but squatting uncomfortably on my bed now, they are impossible to decipher from just millimetres away. Muttering 'full recovery' over and over again, I resume my hard stare, and a number suddenly flickers, then dies away: a 5, and maybe a 3.

Having stood outside silently watching, my daughter now bounds into the room and jumps onto the bed, headbutting me affectionately.

'What'ya doing Mummy? Are you trying to tell the time?' she eyes me curiously.

'I'm looking at the clock,' I reply slowly. 'I'm trying to *see* the time.'

'Oh, well done, Mummy!' she hugs me, oblivious to the subtlety of my reply, and scrambles messily off the bed.

Oranges

Since I came home my walking stick has been my constant companion as I shuffle carefully around the house. Often exhausted and listless, I eavesdrop my family's twittering as they fly nervously around my nest. During the first few days of my return, toy cars, Cindy doll legs and plastic car wheels were swept like snowdrifts out of my way. Rugs, chair legs and side tables were all blockades, perils that were silently removed.

Muffled sniggers to my right warn me that the diligence of those early days is slipping away. Mischievous fingers are inching towards my stick, propped up carefully against my leg. I know that if I don't reach out for it now, it will vanish into a cloud of squeals and a pitter-patter of feet.

Gripping the wooden handle, I prepare to start my old lady shamble towards the kitchen, but something makes me stop. Lifting my stick I gently prod the wooden floor in front of me. *Was that a dim glow down there?* I don't know, but a strong force is making me pause. Something has caught my eye. I laugh cynically at this thought, yet I can't ignore that something has caught my attention. Either way, a sense beyond the act of seeing makes me stop. Spotting my strange behaviour, Ed moves my stick out of the way and, standing back up, produces a tiny piece of withered orange pith, no longer than a thumbnail. Looking at me, he curls the fragment of pith gently into my palm and hugs me hard.

Realisation hits me. The amazement and joy communicated by my husband's embrace was because he thought I had *seen* the orange peel. But life is not so simple – certainly not this life. I understand that while I had indeed known to stop, the message had not come from the obvious place. There was another form of messaging going on – and I am mystified.

Eyebrows

The rota to keep my children's lives running smoothly has spilled out into the local community. A tide of friends flows in and out of my home, collecting and dropping off my children at different times during the week. During one of these exchanges I found myself rooted to the spot. Opening the door to happy shrieks, I was assaulted by shimmering polka dots. Spellbound, I followed the dancing dots in around the mist, clamouring for me to follow them until, like the shrieks, they faded back into the distance.

The welcome aroma of croissants and flowers has permeated the house, as my day has been punctuated with vice-like hugs and encouragement from my relay team of friends. I have slowly hobbled up the driveway several times to delighted claps, and watched my friends gently flirt with my dad and murmur questions when they think I am out of earshot. Alternating between exhaustion and elation, I am in a protective bubble, and there has barely been a moment without a warm hand stroking mine.

When my visitors all finally drift out of the door, the snake is there, waiting for me. I no longer have kind hands to hold onto, and I feel the pull as it starts to drag me under. I am utterly helpless and cannot fight the dread that is building. Panic rises like a tidal wave ready to drown my senses, and I flounder for what to do. I recognise this feeling of helplessness; this *pain*. Yet my breath begins to flow, starting shallowly, but the familiarity and hiss as I release the air slowly out of my mouth is reassuring and audible. I can push back the waves and the water slops down, murky and quiet, but not quite draining away. I can feel the air and the fear being pressed out of my body – for now.

Closing my eyes, I hear Mum poking around the house, banging doors and pressing buttons, keeping the routine moving along. She busies herself by taking on my life, a legitimate usurper wearing her daughter's skin. It's what you do when your child is ill; you peel away the layers of your own life and step into theirs instead. As the machines whirr and the pots boil she is unaware of my silent battle, of my near drowning.

The door bangs and a clatter of feet announces that the children are home. Before I know it wiry legs are climbing up on top of me, bringing the dampness of autumn so close that I can smell it. A serene strength spreads throughout me as cold hands and sharp knees press into my body. My daughter is determined that I will see her eyebrows, and has taken to thrusting her chin up high so I can examine her face. Being fair-haired this would be challenging at the best of times, let alone now, but it is a game we play. She's a child who has never lavished much physical attention on us, so I relish this late development. Squeezing her close I realise that small gingham squares are drifting unsteadily in front of me. My memory knows in minuscule detail what this dress looks like, so this transparent apparition is distinctly unreal. The real dress lies beyond the voile curtain I look through, buried somewhere in my past. Yet the body wriggling inside this dress right now is very real indeed.

Concentrating on the fabric closest to me, warmth radiates up like a haze. This is not just body heat, though – there is a new depth to the murky hues and I start to grin at her.

'Your dress feels red,' I say, and for a moment I feel her stiffen. Yet, this is how it feels! Somehow my brain is trying to tell me this dress is red. My daughter leans forwards.

'Of course it is – *silly.*' Sitting back again, she asks, 'Are you better yet, Mummy?'

'I'm getting better all the time, sweetheart. *You* make me better.' I smile and bury my face in the crimson smells, trying to figure out what this confusing sensory information means.

Later, as Ed takes our son off to bed, my daughter insists that I help feed her blackberry crumble, even though we both know she is old enough to feed herself. It's one of the batches I made this summer, defrosted from the freezer, a relic from 'before'. I not only feel the heat of the steam rising off the crumble, but see a definite white swirl, too. I test the temperature on the tip of my tongue and my stomach cramps at the prospect of food. My daughter wants me to help her, to help myself. I can sense her encouragement and nurturing, and feel conflicting love and guilt. At times my children have a deceptive maturity, an incredible understanding of the world around them. My daughter understands with an innocent wisdom what I need from her, and gives it without a thought.

Voice commands

I am seeking out a new independence, and technology is for once being my friend. Ed guided my finger to a button at the bottom of my Blackberry earlier today, so now I can voice command it to call my friends. The wondrous feeling of being able to call whomever I wish, whenever I wish, is amazing. The only downside, of course, is that people can call me back. The shrill ring now evokes panic as I realise I will have to answer it myself. I am hesitant as I cannot read the screen or easily find the button again, so while I have some independence back, it has created a new set of problems.

I have also taken to writing notes, but it is a strange experience, writing without seeing. I have to measure the spacing with my thumb and form the letters carefully, only writing using a jumbo felt pen. Even then the lines are so faint that I have to squint and hold the paper up close to see them – the acrid fumes of solvent pervading my nostrils. The worst part of it is all the damn helpers tidying up my precious notes. I cannot leave one out on the arm of the sofa without it being swept away onto an invisible pile. Ed sighs and digs the notes back out of the mountains of paperwork that litter our house, and wordlessly hands them to me. Anger and frustration at my pathetic attempts at freedom pour out of me, as the very people trying to help me are the ones thwarting me.

Laundry bag

It's a grain-of-sand day. I don't yet know what the changes are, but something is different. There's some kind of new perspective emerging through the fog. My ritualistic scan of the room instantly drew me to dark lines where the gaps between the wardrobes should be. They weren't there before, or certainly the lines weren't so visible.

The lines flicker like a vintage film, and it's unsettling to focus on such a jumpy and erratic picture. But, there is definitely something *new* today; and that makes it a grain-of-sand day. Even this imperceptible shift makes me squeeze my eyes shut and whisper a silent thank you. I don't really know who I am thanking, but my eyes are full of grateful tears.

As I settle back onto the pillow wiping my cheeks, I catch a smeary glimpse of the wicker bin that lives in the corner of our room. I know wicker has a weave to it, but maybe I can see just a little of that detail today? Okay, so it's bin shaped, I know that, but I am sure there is a texture there too – so faint it's hardly perceptible. But something is making me stare at that wicker basket, and I'm learning not to ignore these impulses, because they normally mean something. Tears start to drip down my cheeks anew. Who would have thought a waste-paper basket could be so marvellous?

I hear the children bounce downstairs with Dad, and the day erupts into life. I stay in bed, tired again. The ghost of unease is skulking its way across the room now, its shadow snaking across my bed, stretching up its arms to me. My initial peak of joy is dulled by the reality of how far I have yet to go. I struggle out of bed before the spectre can ensnare me, and wobble over to an old Ikea bag that has been abandoned at the foot of the bed. I know these bags are made of crinkly blue plastic – the filing cabinet in my brain holds that information. However, what I see now is paper-thin and indefinite. This morning, light bounces off the plastic coating, shooting dazzling white beams at me. Folding my stiff body in half, I start sifting through the clean laundry inside.

Confused messages bombard my brain; this act is both familiar and foreign. Pyjama bottoms and T-shirts feel wonderfully soft and evocative, yet as I hold them up my eyes don't register them. It is only when pressing a tiny vest so close to my face that I can smell the remnants of fabric conditioner, that I detect a tiny pattern. Little piles of clothes are getting larger; my hands are my eyes, identifying zips and buttons. A voice deep inside me growls a warning to the ghoul that is still sliding around the room. *Go away*, I mouth silently. *I couldn't do this if I wasn't getting better. Can't you see I'm sorting out the laundry?*

Weighing things up

I had a dream last night. I can retell it all, because like everything else, I narrated the details of it into the MP3 player I now carry everywhere with me. But, as is the nature of dreams, it has since evaporated from my memory. It was a harrowing vision and one I thought I couldn't possibly forget, but my mind seems to dissolve things that serve no useful purpose. This in itself is surreal. I no longer have the ability to remember the dream, yet its ethereal shadow has been given an electronic reality it was never meant to have.

My dream was very simple – my sight came back. While I'm sure this is common for those who experience sight loss, what shocked me was the disinterest of friends and family. In my dream no one seemed to be remotely interested, and the sense of loss and confusion when I woke up was palpable.

Later, as our daughter climbed into bed with me, she announced in her matter-of-fact voice that I looked *flat*. Looking down at what part of my thinning frame I could see, that morning emphasised by shapeless grey pyjamas, I had to agree with her. I am flat. My body looks deflated; my illness has sucked all the life out of it. The dishwater hue of my body and this lack of humanness about my physical self doesn't help the sense of nothingness I feel. I don't look alive, so how can I feel alive?

As the children clatter out of the door with Ed, I lumber into the bathroom, my daughter's comment reverberating around my head. The house seems eerie and I instantly miss the physical nearness of my husband, standing as he so often does, silently behind me. Shaking my head I decide it's time to know how much weight I have lost. However, in order to read the glass scales that I prod tentatively with my toe, I realise I will have to employ some daring tactics.

Initially standing on the scales, I sigh as there is no way I can see my feet, let alone the screen beneath them. Stepping on again I attempt to lower myself to the floor, but I am slow and the electronic dial soon flicks off. Forcing rubbery legs to kneel and with my nose just centimetres from the dial, I press the scales again and can almost make out one faint digit. Exhaustion overwhelms me and for a moment my head slumps onto the glass. As I lie there feeling utterly hopeless, my nose suddenly twitches. The scales smell of talc, yet as far as I know no one in our house uses it. This pleasant yet alien smell galvanises me into action. I decide I want my house free from the strange bodies that inhabit it. I want my life back, and that means finding a way to overcome this challenge.

Experimenting again, I work out that after pressing the scales I have around three seconds to read the dial. This exercise will require speed. Thankful for once that the house is empty and my family are unaware of my circus antics, I practise an inelegant 'hop-off' technique. But, I am just not fast enough. Unfortunately, I need to hold the wall for support and that slows me down. I try kneeling my weight on the scales, but that only results in me headbutting the toilet bowl. Tumbling is taking on a whole new meaning in our upstairs bathroom. All dignity now lost, I crash to the floor in a magnificent bungee-style descent that leaves me spreadeagled, but able to just glimpse the numbers before they blink off.

Unfortunately I am required to do this painful manoeuvre numerous times to read *all* of the numbers – but it works.

Not surprisingly, my weight has decreased. The dial registered a perverse number that would normally have left me excited at the prospect of a smaller dress size. However, right now this is a deep concern. I am eating as much as I physically can, but my stomach scrunches up as soon as food is even mentioned. Mum surreptitiously leaves bowls of nuts next to me, but even nibbling four cashews is an effort. My relationship with food, once so unquestioned, has become a battle of need. I had no idea my sight was such a part of that relationship. Invisible food tastes – invisible.

This situation isn't helped by the jovial and sometimes careless remarks that some visitors make. Complements on my bagging waistband and sharpened cheekbones make me recoil in shock. Sometimes I feel that my visitors are missing the true severity of my situation. On the one occasion when I sensed that my reduced size had evoked a misplaced envy, I resorted to extreme means. A quick retort suggesting that my visitor and I swap places halted any further remarks; but like any invasive treatment it also killed the conversation. I found myself remorseful and contrite, aware that I cannot blame others for what has happened. I know I must tread carefully, for it is these thoughts that become snake fodder.

My weight has been discussed, and concerns have been raised at the amount I've lost. But even though I am helpless on the outside and reliant on those around me, I have secrets. I may not see much, but there are things they don't know about me, too. For while we talked in detail and devised a food plan to help me gain weight, at no time did anyone think to ask me how I read the weighing scales in the first place.

Dusty corners

At times negativity pervades my thinking, and it's hard not to want the visual changes to be more obvious and more frequent. I try to rejoice in small accomplishments, but it saps my energy. My success rate for identifying red has increased substantially, and I am now correct around 80 per cent of the time. Some muted reds still elude me, but the bright bold ones shout out and assault my senses. Blues are emerging, too, sneaking into my consciousness; not just a tone – a playful hint – but a tangible colour. My family listens to my constant dialogue with curiosity, hearing but not understanding this new language of seeing. They have no idea what I see, or that my emerging world is no more than dirty water. They don't understand that even though colour is warming my senses once more, it's not colour as they know it. The difference now is simply that I know a colour is there, but it is only the shell of what I once knew it to be.

I am endlessly counting, waiting and assessing every blink. I scan the room obsessively, I mean, really *staring*, straining to see anything – something. The smog floats close to my face, a grey veil I am forced to wear. It's dark, too – not that murky, middle-of-the-night kind of dark, but a brown dusk. I constantly demand that the lights be turned on, even on sunny days. Even the anglepoise lamps Ed has bought don't seem to help. I pad around positioning the lamps around the room, while Ed silently follows me around, turning them off.

I try days without my glasses on, for they do not sharpen this world. What would normally be a lifeline is now a hindrance causing the lines I do see to jiggle all the more. Yet when I take off my glasses I instantly miss their protective weight on the bridge of my nose. Sitting on the sofa, my glasses in my hand, I'm suddenly aware of the criss-cross base of the table next to me. It has just popped up like a 3D card! I don't know if these tiny improvements happen all the time, or whether it takes me a whole day to acknowledge them. Recently dark, dusty corners are starting to reveal secrets and previously hidden treats. My life is reappearing slowly around me, like a magic show in reverse. I can locate the banisters that Ed and his brother installed when I first came home, and I know they have knots in the wood. Of course, I don't *see* them as everyone else does, but I can feel their knobbly imperfections as my hands grip them. No one else cares, but there are places where the smoothness is broken and the life of the wood shows through, and my hands find these secret flaws.

Ed is tired; he is juggling a job, looking after me and shouldering the stress of the unknown and all the fears that serious illness brings a family. There are times when our individual worries manifest and collide in a night of turbulent sleep. He has to divide his time and energy between two energetic children and me, and we are all sapping him. I can't help it, but every moment in this unsettling world makes me crave reassurances I would never normally need. I can't see it, but I can sense Ed's fatigue for I am a sponge soaking up his worry – his worry about our future.

I move to camp out on our chaise longue next to the patio doors, ignoring the cold that seeps through the glass. My attention is continually drawn outside, craving sunlight. I am a moth seeking out the brightest spot. My visual range is still only a few feet, but I can make out shapes and silhouettes in the garden, and it is all reassuringly familiar. I long for more, much more; this visual rebirth is taking a very long time.

Mr Remote

My obsession with inanimate objects has continued now that I'm home. It's not just that I need to register objects close up, but some items have sparked what I can only describe as compulsive behaviour. I have discovered that if I hold my dictaphone right up to my face and tilt it slightly to the right, my peripheral vision seems to be a fraction clearer. It's still cloudy, of course, and the detail just isn't there, but if I wiggle it a little, I wonder if there is a flash of clarity, of *more*. I scrutinise it for hours, tilting it one way then another, straining to see a button or an edge. My daughter eyes the device and me suspiciously. I try to laugh off my strange behavioural ticks as *Mummy being silly* because I don't want her to worry. It's important that she doesn't know the full ramifications of what has happened to me. I know she is always watching, and I can feel her five-year-old eyes study me with an astuteness they should not have. I gently twist my wedding ring round and round – that's become another habit. I daren't look down because I know that if I do, I won't see my ring. It's so peculiar not being able to connect with my own body. I can touch my knees and legs, but they hover out of my visual range. I am a limbless torso covered by a milky blanket and I long to be whole again, to be a mother to this little girl staring at me.

When boredom finally forces my daughter from the room, I reach for the remote control and scrutinise this instead. It feels heavy and awkward in my numb fingers. Not surprisingly, I've never smelled the remote control before, but I get whiffs of plastic and Ribena. The buttons tease me, gently backlit, but just not light enough for me to identify their purpose. I find myself pointing the remote tentatively at the TV as I try to connect the radio through Freeview. I hit the wrong button and instantly lose any idea of what is on the screen. Panic rises in me as a dull fizzing sound emanates from the television, and I feel utterly helpless. I call Ed frantically, overcome with an irrational self-loathing. I hate feeling so physically destitute, so clumsy and uncoordinated. It's just not *me*. I am the sharp one in this family. I'm the one who changes the towels and descales the kettle without anyone noticing. I'm on the ball – I make things happen. I'm the most *make things happen* person I know, yet right now I can't even turn the damn television on.

'Lellow

Dad has gone out to the park with a local girl called Connie. We've had several girls come to see me, and my family refer to them as 'options'. We've tried childminders, mother's helps and nannies, and an au pair was mooted, too (but instantly rejected by Ed). This girl is only sixteen, but the children seemed to warm to her instantly. The plan is for Dad to let her lead the way, and his job is to see how they all get on. I'm happy to let them go, relieved that I no longer have to play the role of invalid in the corner. I think about that word as I repeat it into the dictaphone after they've gone. *Invalid*. That is how I feel, as though I have no use. I am somehow not *valid* anymore.

The Hair Flicker has organised help in the mornings, and an invisible woman distributes milky cereal and ferries my children to their childcare. Familiar sounds float up the stairs, but they are followed by unfamiliar voices. The routine is fluid, changing each day, and its unpredictability is a cause of unease. No one knows how much care I will need, or for how long. It is proving impossible to bridge all the gaps and I can't see the human traffic jam easing anytime soon.

Fingering the dictaphone I demand that my brain sees the details on the front. The heavy weight tells me that it is metal, and I feel the straight edges and the raised bumps of buttons, but other than that I cannot make out the model or any of the lettering. I don't let this dishearten me and, leaning on my stick, hobble over to gaze up at the clock in the kitchen. Just recently I have been able to see faint arms emerge through the haze. I remind myself that it may take many minutes of concentration, but eventually I will read the time, and that means I can control one part of my life again.

Later, when the children are back, my son saunters over to me with carrot sticks balanced on a plastic plate. We cuddle quietly on the sofa, the only sound being his comedy crunching. As his little body starts to relax into mine, his plate, which had been resting on my knee, falls to the floor. Before I know it I have jumped up, shrieking.

'Yellow, it's yellow!'

Dad sticks his head out of the kitchen and asks what's going on, but I just burst into tears. The plate is not a tinge and it doesn't just *feel* like a colour, it *is* yellow. This unpredictable and confusing visual journey has just thrown another miracle at me. My brain circuitry has allowed me to see the first yellow thing in weeks. As I sob on my hands and knees picking up carrot sticks, my two-year-old son hands me the plate, grinning.

'Is 'lellow, Mummy.'

Washing machine

Determined today to be of some use around the house, I decide to put the washing on. *The Hair Flicker*'s helper has filled it up, so it would appear that all that is left to do is to turn it on. However, this means that I need to be able to decipher the settings – no easy feat. Once again I am aware of other fingers twisting knobs, rearranging things, subtly resetting my life.

Even crouching down is tricky as my feet are still unyielding, and balancing any weight on them is a challenge. The sensations in my feet have become more and more bizarre of late; my toes now feel as if they are stuck in gritty quicksand. It is very surreal, and I want to wipe them clean, even though they are spotless inside my socks. Kneading the pads of my feet, I plead with them to start feeling like proper feet again.

Taking off my glasses, I peer at the black squiggles on the washing machine. Helpfully, one setting is bolded out, so I aim the dial round to here. As the water starts to gush, I feel a wave of simple pleasure that seconds later is replaced with feelings of utter misery. If putting on the washing machine requires such concentration, how will I ever do all the other chores around the house? How can these simple and thoughtless tasks become life-changing events overnight?

I sigh as emotion threatens to overcome me, and stagger up to my feet muttering *full recovery, full recovery*. My daughter's latest painting is casually hung over the corner of the chalkboard. Made up of block colours and a childishly vivid face, it has repeatedly grabbed my attention. I'm told it has her inimitable style of overly large eyes framed by comedy eyelashes and a wide, smiley red mouth. Today, I know that the face is framed with wonderful lemon-yellow hair, and I am 100 per cent sure that those colours are real. As I stare harder the red lips become stronger and more intense, eventually blowing a kiss at me. I am taken aback. The colour intensifies the longer I stare at it and I wonder if my brain reads the scene, but then needs time to deliver the message. *Red? Are you red? Yes, we're red!* The lips scream back at me and suddenly I'm giddy with joy. Today there was no guesswork at all; red has found me again.

Falafel lasagne

Family visits are becoming more frequent. My sister-in-law Su and her entourage are coming and I can't help but feel like a zoo exhibit once more. I can already imagine my nephews cocking their heads and examining me with naive curiosity. I am on guard, nervous, and the snake is stirring.

I know they are coming out of genuine concern, wanting to offer any help they can, but it frustrates me that I cannot control how I will feel when they are here. I wonder if Ed believes these visits will make life normal again. No one seems to understand how stomach wrenching it is to have anyone (even those I know well) come into my home, my safe place. No one seems to get how much things have changed. All the visual cues I would have at my disposal are gone. I cannot do the invisible scan that sighted human beings do on meeting one another. The ability to assess and absorb information is so imperceptible, so natural and so unthinking that my family don't even realise they have it. How do you explain to someone that you have lost the very thing they don't even know they have?

Before I know it, feet clatter through the door and muffled hellos are accompanied by the sound of shoes being kicked off. Voices suddenly mute unnaturally, and I hear my two nephews quickly diverted into the playroom. My senses are twitching, my hearing on full volume, and I hear Su pause as she approaches me for a hug. She smells of shampoo and cigarettes, and her outline is familiar yet ghostlike. In true fashion she makes cheerily inappropriate comments, but I can feel her warmth radiating through me and she feels real. She chastises me for my ratty nails as the multicoloured varnish is now badly chipped. Offering to take it off, she disappears to find my make-up bag. At the same time the men gather up the children and tumble back out of the house to take them for a walk in the rain.

While an impromptu manicure is a welcome distraction, Su and I are both out of our comfort zones. But I like the feel of Su's hands, that physical connection again that I crave like a drug. If I hold onto another human being I can't float off into the mist and dissolve. My nails feel wet and uncomfortable, causing the snake to uncoil, and anxiety ripples through my gut. I ride the waves, breathing over them, soothing the tide back down. The dark water subsides enough for me to talk, to spill out the few things I was able to do today, the washing machine, the laundry. Silly little things, and once again I am the child being gently encouraged and praised.

My visitors have brought an early supper with them – a lasagne. This has been a common occurrence since I got home, the *Bringing of Food*. As the cooking aromas waft into the living room, Su cheerfully announces that she chucked everything from her fridge into the lasagne; including falafel. At the mention of this my insides roll yet again. The thought of eating a heavy meal is already making me sweat, but to be told it is a falafel lasagne almost undoes me. A lifetime ago this would have been funny and an opportunity for silly jokes. Today my sense of humour has long gone and the thought of any food is torture.

At dinner I force down tiny morsels and eventually the day draws to an uneasy end. At her cousins' departure our daughter erupts into an ear-splitting tantrum, her disembodied screams making her sound like a captured animal. Writhing and kicking out at Ed, she is carried off to bed, and I am left feeling utterly helpless. I don't have the emotional capacity to deal with my child's outbursts, her angst. I understand her fears and her frustrations, but I can barely cope with my own emotional levels. I am already spilling over.

The tension is still lingering the next morning. My stomach lurches with nausea as the strain in the house is like treacle, sticking to all of us. I realise even having people helping us is a sap on our energy. I need to escape from this tautness in my chest. I can't breathe properly. I need to switch off, to flick a switch and just *not see* for a little while. I am reminded of my sight constantly. I can forget my feet and hands, but cannot ignore my sight as I peer through this grotesque porthole every waking second. I long for the moment when I can retreat upstairs to my tropical hideaway and escape this awful place.

I finally told Ed about the snake today. I realised part of its power came from its invisibility. I know I need to learn to train it, to avoid provoking it. It feels as though all laughter has left our home. We are the dark house on this street, the one with no fun, no light – and our children know this.

Of course we didn't know it then, but humour would creep back into our lives and we would all be able to laugh freely again. One of those moments would be when Ed discovered the second lasagne that Su kindly brought that day, buried deep inside the freezer. She had scrawled '180 degrees, 40-min. lasagne' cheerily on the lid. We laughed at the memory of its secret ingredient, and marvelled at how tasty it was the second time around.

Recycling

Our front door bangs shut so I creep silently into the quiet kitchen. I can still feel the imprint of my children lingering in the air. My leg knocks the recycling bag left littering the floor, and I realise with irritation that it is collection day. Frustration that someone has forgotten to finish this mundane task rises up and threatens to overtake me. I lug the heavy bag down the hallway swallowing back angry tears. On the chilly driveway I have to kneel down to get my face close enough to identify the recycling boxes. Pebbles dig into my knees and nettles in the cracks of the driveway sting my forearms. I know one box is a dull blue and the other a dirty green, but through the mist they just jeer back at me. I have no hope of identifying muted colours, for these soft tones have not yet stepped back into my life; it is only the brightest and loudest colours that are talking to me.

Stale wine, like vinegar, pricks my nose and I realise that one box smells more tartly than the other, so I make a choice. Item by item I sort the plastics and paper into their corresponding boxes, cursing each sour-milk carton and sticky jar, refusing to let discarded objects defeat me. My mutterings turn into words that get stronger and louder as my anger becomes a kind of strength, easing my frustration and driving me on. Yet again I am stunned that such a simple task is so extraordinarily challenging. Looking around, the grey mist wraps itself around my head, obscuring everything. I can only see a few feet away so my garden is just a blur of dark shapes, a void. *Sod off mist!* I hear myself say out loud, *I want my garden back.*

By late morning my sight has started to settle down and the mist has thinned out a little. I am becoming more conscious that the whites (or at least the whitest shades I see) are overly luminous, and imperceptibly flashing, like an erratic strobe light. This world I see is volatile and unhinged, and at times like this very disturbing. I am trapped inside a black-and-white kaleidoscope, mesmerised by the fragmented view that rotates in front of me. I am exposed to the very apparatus that makes up the basis of my sight, and I don't think we are meant to see this behind-the-scenes rigging. I am suspended in a halfway place, a prechromatic land, as my visual system is slowly initiating what is to become my second sight.

Almost without me knowing it, the cooker clock has blinked its way back into my life. My daily test of the illuminated numbers has yielded varying results. Sometimes I can read them during one part of the day, but then not again later on. I try not to get disheartened and remind myself that I have come from complete blackness; that there has been improvement. It might be that a stranger walking into our house would think me a simpleton, or maybe just plain crazy – a pale-faced, black-eyed woman shuffling around whispering and touching chair backs, table legs and door frames. But I know that I am triggering the swollen circuitry in my brain, nudging it awake again, reminding it of how the world should look. I can't explain it to others, but I know that what I am doing is not madness; it is driven by a deep-rooted wisdom that even I do not understand.

There have been muttered phone discussions going on all day between Dad and Mum. They are planning a handover, but I am sensing that there will be a gap and I'm filled with dread that I might be left to my own devices. They don't seem to understand how monumental that would be. I start planning who could visit so I won't be left alone, not even for a minute. I don't want any opportunity for my snake to be disturbed, to start uncoiling. I fear something terrible might happen. I don't know what would happen, but it is very real, and all I know is that it would be very bad. A loud, buzzing noise jolts me back to reality. Heart pounding, I flap around trying to press a key that will quieten the metal bee next to me. A tinny voice tells me he's calling from the training company I had contacted months back about an executive coaching course. He wants to know if I am going to enrol onto their diploma. I hadn't quite got as far as paying the course fees and memories flood back as I realise my old life has just called me up. Suddenly that goal feels so far away, so unattainable. The choices I wanted to make back then now feel about as worthless as the debris I have just sorted through outside.

I need to do something. When the grim reality of my situation starts to override my senses I know I need to counteract quickly. I pick up my phone and using voice dial command it to call the Frock Exchange, a local shop that sells secondhand clothing. I chat to the lady easily, joking with her that making £6 from selling my children's old clothes is better than nothing. It's a commonplace conversation, nothing untoward or even that interesting, yet the experience is surreal. It is *me* talking – yet it is not me. I am acutely aware that she has no idea of my situation, of what has happened. My voice might be vaguely familiar to her, but she knows nothing more. Part of me wants to shout down the phone, to scream *Don't you know? I can't see!* But, part of me is proud for keeping it together, for holding in any outburst. It shocks me that we can live our lives without really knowing what those around us are experiencing; even those on the other end of a phone.

Touching the real world has left me with a sense of shame, a bitter taste in my mouth. I do not want people to know – I don't want to admit to this illness. I guess it must be a form of denial, but it goes deeper than that. I don't want to be disabled, to no longer be of *able body*. I don't want to feel less of a person. Yet right now – I do.

CHAPTER SIX

Mowbray Road

I have many regular visitors, but one local friend visits me daily. As I hear Genet knock gently on the door, I reach for my coat. We have been friends for several years, our daughters sharing a strong bond from nursery, and I look forward to these walks. If we turn right at the top of my driveway, my slow shuffle eventually leads us to Mowbray Road. The sign locating this road has been my nemesis from the day I made it to the top of the drive. I stand before it, straddled like a cowboy with my hands on my hips, my stick propped against the wall. *I will read the words.*

It took more than a week for the black letters to untangle themselves and become legible. Today I can slowly read the sign from around four feet away, and it is a new benchmark. I am no longer rubbing noses with cars and lamp posts; I can make out letter shapes from a more stately distance now.

Along with this visual progress there is a new development in my limbs. Replacing the sensation of grit, my feet now burn and tingle. The numbness is melting away and I can feel fine sand between my toes, replacing the coarse gravel and quicksand of a few days ago. They still feel awkward and restricted, but this is slackening off, too. I feel a smile twitch as I consider the wide range of garden aggregates available on the market, and I wonder if my brain is going to go through the lot.

Smiling hosepipes

I'm standing in the garden and cold air creeps around me like a damp cloak. The grass is spongy and slightly wet under my feet. I am here to experiment with what I can see today, and I will know this by listening out for what objects speak to me.

My head jerks to the right as the hosepipe, left casually strewn on the lawn, stirs and smiles up at me. 'You are yellow,' I whisper conspiratorially, and the hose curls its tail in delight and glows up at me.

Taking a step forwards I nearly tread on a wet plastic ball starting to disappear into the overgrown lawn. This must be one that my son chucked outside months ago, another remnant from the past. I stare nostalgically at the round shape for long minutes, not daring to ask myself what colour it is. It feels murky, mystical and unfathomable, and a thousand thoughts fly through my mind. Memories of colour pop up, but nothing matches what is here. A lifetime of sight tells me that grass is green and skies are blue, but even that certainty, that absolute knowledge, is not enough. My brain is processing mud.

I feel a battle going on inside my head. Education, knowledge and experience tell me that I am surrounded by colour, but my mind just cannot decipher it. I know I sense red as warmth, an intense sensation rather than a certainty, and something in that ball is *red*, yet it is not. None of this even makes sense to me, so how can I explain this incredible mismatch of information to anyone else? As the cold air twists its icy fingers around my ankles, I realise I will have to puzzle the world back together by myself.

Eventually, after flicking my gaze back and forth between the grass and the ball, I decide that the ball is in fact green and pick it up. Still studying it closely, I limp back into the living room. Dad starts to speak but I interrupt him, eager for the answer.

'Is this ball green?'

'Yes.' I am close enough to know he is smiling at me, albeit slightly bemused. He knows this answer will not be sufficient, so he adds, 'It's bright grass green.'

I gasp and clap my hands in delight before I freeze. 'You didn't see it on the lawn, did you?' I ask, staring back outside, realisation dawning on me. 'None of you did. You all missed it?'

It's a hypothetical question, of course. The ball is green, so it's not surprising that any normally sighted person would overlook it. A green ball on a green lawn – it's the perfect hiding place. Of course, the other critical factor is that none of my family was looking for it – they were not expecting it to be there. They do not scrutinise and pick apart their world in the way that I feel driven to these days. It is only because greens have been intermittently flashing up as red that I was able to see it myself. The irony is both exhilarating and simultaneously crushing, for it is the very faulty nature of my vision that has allowed this soulless victory.

Life drawing

Someone has been drawing inside my head. My teenage self, obsessed with Hockney's line drawings, is hunched over with a fine liner busy scrawling deep inside my cerebrum. Thin black lines appear almost daily on my horizon; the missing detail from my life is being drawn back in. It's as though my brain has got hold of an Etch a Sketch. Unfathomable matter that was not visible through the murky mist yesterday has today developed outlines and shapes.

A line of cars parked on either side of Mowbray Road stops midway down, simply disappearing into the swirling vapour beyond. There is a parapet, a boundary to the world I see, and edges fall away into nothingness. I do not have the individual words to describe this; it's like the transparent part of a negative where no visual information has been recorded – an ocular hole. Yet today there is something there; for here is a thin grey line forming the shape of a new car magically appearing on either side of the road. A surge of joy threatens to spill out as I quickly count the houses down each side to the point at which they dissolve, and cheer when I can count one more. My landscape has just extended by the length of one car, and I feel my world shyly starting to reveal itself to me again.

Jogger

Ed and I go for a walk at South Norwood Lakes, which is walking distance from our home. We are managing one whole circuit now, which is exciting progress, and I am starting to feel the wind in my hair again. The physical changes are so much easier to spot; I don't even have to tell Ed, for he can see them for himself. My vision is harder to express. I try to explain that I see contrast; that I see differences, what is *not* there, rather than what is there. Faces are the hardest to negotiate visually; they are delicate and indistinct. I see eyes the best, the whiteness often contrasting with the iris, giving me a division – a stark contrast. An iris is an indicator that there is something else there, a living body in among the mist.

I cannot choose how my sight returns or how my perception of the world manifests, but the baffled sighs around me express a deep frustration and confusion. As we walk I point out road signs that are starting to produce an iridescent glow on the horizon. Caught inside the vacuum a new outline grows, a perfect rectangle – an identifiable object.

As the rhythmic pad of a jogger comes up behind us, Ed carefully sidesteps us out of the way. A flash of red, or perhaps pink, laughs back at me and leaves me flummoxed. My mouth gapes and Ed squeezes my hand in concern. Turning to look at him I see his face, *his body*. In the natural light he has filled in and is more whole than he has been in...forever.

'I saw – *pink?*' I stumble. 'It literally slapped me in the face!'

'She's wearing neon,' he replies, his hand transmitting a smile. I squeeze back, and swinging our arms enquire playfully, 'Was she Asian?'

'Wow!' he exclaims, turning to me. 'Could you see that? Yes, she was, Chinese maybe.'

I didn't see her face, but I did see a black waterfall of hair and something angular about her jaw. I made a best guess. She was still a see-through apparition, but I am starting to recognise ghosts now and they do not all look the same.

More spooks visit me, recognisable voices and familiar laughs fill the living room, and I sink into the social hum. I am quietly revelling in the acute pleasure of having just made my own cup of hot chocolate. It sits cupped in my hands, on display for everyone to see, even though they don't know that.

I had to overcome my fear of the kettle to make this drink, and I am still feeling wary of it. It boiled furiously and hissed steam as it sat regally on the countertop. I couldn't separate its metallic edges from the dark granite, so I had to wait until its angry bubbling had subsided before I could get close enough to pour the invisible liquid. It slopped over the edges of my mug, but I didn't mind.

As I sit listening to the drone of voices around me, my mind wanders to the urine-soaked underpants that I know must be still on the bathroom floor upstairs. My son had an accident earlier today and I haven't heard anyone sort out the mess yet. He was ushered up to bed screaming with all his might, and the house shook with his pint-sized anger. Unfortunately, the house is full of men this week, and I know sodden underpants are low on their priority list. There was no dinner in the fridge for the children this afternoon either, and pasta was hastily thrown together. The men are oblivious to these details, seemingly unimportant household irritations. Yet these are the small details that I know about, and they frustrate me enormously. My inability to stride over and just sort them out is making my chest tighten and constrict.

I breathe out slowly, undoing the belt across my chest, and gaze at the face nearest to me. I am happy staring at this one as it offers a smorgasbord of detail the other ghosts cannot. Dark skin provides magnificent contrast, and eyes pop out like beacons. Even a faint sheen of moisture is visible as the light paints stripes across her face – stripes that I can see. Her smile grows into a guttural laugh, and I revel in whiter than white teeth that smile right out at me.

Cable ties

Ed is supportive, caring, sometimes distracted and occasionally provocative. As our walks are becoming longer and more regular now, he has taken to challenging me over what new colours I can see. This is not as simple as it may sound, for we do not see the world in the same way. It's something we are both aware of, often finding that a disagreement was simply both of us believing the same thing, but coming at it from opposing perspectives. Identifying colour, not surprisingly, also falls into this category. In order for him to test me successfully we would need to define colour in the same way; and we don't. Unfortunately my turquoise is his green.

Walking past a corrugated plastic gate held together by cable ties I find myself pausing, drawn to this seemingly mundane scene. To me, however, it is these very cracks of life, the ordinary and dreary scraps that are feeding my brain, filling in the spaces where I know colour belongs.

Our conversation starts with the basics – the colour of the corrugated plastic itself.

'I'm pretty sure it's grey,' I say, bending awkwardly to run my fingers over its crinkles and bumps.

'Nope,' Ed quickly responds.

'Well, it is grey, I'm sure,' I reply, standing back up. 'Okay – the sun is on it and that's making it appear grey. I know that but I don't register it as black.'

'Ah, well, yes, the sun has bleached it out, but it is in fact black.'

'Yes, but that's the point, it appears grey to me right now. So therefore it is grey.'

'But, it *is* black.'

'What colour it is without the light on it doesn't matter to me right now. If you got a pantone reference and put it next to it, the range would be in the greys.'

'Okay, whatever; but it is black.'

These conversations are frustrating for both of us, but persevering, Ed suggests we move on to some cable ties holding the gates together. At these words I suddenly become aware of previously invisible plastic ribbons, neatly knitting the edges of the corrugated sheets together. Surprise silences me, and Ed gently takes my hand as my mouth drops open.

'What colour is the top one?' he asks softly.

'Hmm,' I reply, for I have only just begun to make out the elusive tie now that he has directed my attention to it. 'Red?' I guess wildly, still puzzled.

'Nope, its...' he pauses infuriatingly for a beat, 'yellow.'

As I hear his words my brain syncronises this conflicting information and I find myself believing my ears, and the top cable tie glows and jiggles in glee. I am dumbfounded – and shocked enough to take a step back, almost tumbling into the road. I don't need to tell Ed what has just happened; my reaction tells him everything.

Stunned but curious we work our way down the rest of the cable ties. My brain is unable to process the colour by visual stimuli alone – it needs Ed saying the colours out loud before it then magically lights up. At the end of our test I can see most of the cable ties with their correct colours, but like dying embers they quickly fade away. Ed's voice reignites them again, making them flicker back to life, teasing me. We pass this gate every day for the next month until the plastic sheets are finally replaced with proper garage doors and my playful experiment disappears forever.

Spiders

There's a soft warmth next to me as I wake up, a small feral animal still sleepy and surprisingly tactile. My son gently strokes my face and whispers toddler gibberish into my ear. Slowly opening my eyes I am shocked as spiders spill over the room. Tiny black arachnids infest my vision, crawling over everything around me. This isn't fleeting and I have time to study the horrific mirage as it takes over the whole room. As I look down the spiders crawl over the sheets, over my son's skin – and I realise I am holding my breath. Blinking frantically, I pray this is a fantasy that will pass – yet it feels so incredibly real. A few minutes later the infestation has ebbed away and my brain has cleared. I look tentatively back down at my son – the spiders have gone, but the plague is still very present in my mind. As he gazes innocently back at me I can see the edge of his iris. There is no colour or detail, yet I can gaze deep into his soul. It's amazing how little sight you need, to really *see*. He stares back at me, utterly guileless, and his blue crystals send a volt of energy through me.

The spiders stay with me for much of the day, creeping somewhere inside my brain, hiding away until they are next disturbed. We can attach no logical explanation to their sudden arrival, and accept that this is simply another visual nightmare I must endure along this path of recovery.

Even at two years old my son is the house comedian, and he performs for us later in the day. Sitting precariously balanced on the toilet at bath time, he senses he has his family's full attention. His monkey antics and random babbling escalate to the point that he almost topples backwards into the toilet bowl. Darting movements and the sound of shrieking laughter echo around the room, and I can hear the sounds of a normal family doing normal things. I smile, too, but mine is an invisible moment, for my connection is almost entirely auditory. I am projected to the outer reaches of the room, lonely and detached, just my secret spiders and me. My family don't know that the startlingly bright bathroom spotlights burn out any detail I might have. I am separated visually from them, only hearing what is going on. My smile is not because of what they see; it is because of the laughs I can hear.

The day has been busy and the snake has been active. It is a difficult balance of doing enough to be occupied, but remembering I am still ill and need rest. The moment I do too much or go too long without a break, my whole body starts to jitter. This evening the only way we can make the shaking subside is for my family to wrap me up in their big, tangled embrace. I feel the weight of their warm, clean bodies lying quietly as arms and legs intertwine to make one single, breathing being. The energy flows and I breathe deeply again, transporting us all to my empty beach – and for a little while I feel strong.

The blue room

We are back outside that small, stuffy room with the sticky electrode machine that measures my brain activity. Except this time I am walking (well, hobbling) into the room on my own two feet. The first time I was here I saw nothing, my entire body being sucked inside an endless black abyss. But this time it's different – I can see.[*]

I am greeted by a shaggy-haired man who as soon as he speaks I realise is my technician friend from my first visit here. We both know that back then I couldn't see anything at all. *Shaggy Man* takes my hand and smiles, and uncertainty washes over me. Suddenly I understand that he is shocked that I am walking, that I can see him – that I am here. I hear the 'miracle' word sprinkled in the air again, and my discomfort rises further. The snake has come with us today, gently slithering around my intestines. I feel it nudge the wall of my stomach and I lurch slightly to one side. My friend guides me into the room and I stop still in shock. It is blue. Well, not entirely blue, but there are unfashionable cupboards (probably circa 1988) with round pine handles lining the back wall. I scan the room, tears pricking my eyes, my confusion starting to make a fizzing noise in my ears. *I know this place, yet I have never seen it before.* I reach out and touch a TV screen, hard and metallic, and note that it is on wheels. There is the chair Mum must have sat on behind me, and as I turn my head all the way back around I see a random coat hook on the wall right where I had said it was.

Thankfully, on this third visit I can just make out the screen with its checkerboard pattern. So after *Shaggy Man* has

---

[*] When my VEP (visual evoked potentials) were first recorded, a large strobe lamp was used, flashing an intense white light once every second at my unseeing eyes. These lights are normally reserved for infants or unconscious patients, but as I found out they can also be used with blind patients.

glued the electrodes onto my head, once again the VEP
machine starts its beeping and we wait nervously.*

The snake allows me to take the test again. It allows me to
nod and thank my shaggy-haired friend and hurry out of the
room. Then it strikes, hard and fast, and I am unable to
breathe. Ed is strolling – he wants a coffee in the cafe – but I
am staggering out of the hospital door so fast that he has to
follow me. He can't see the snake, but it is coiling up around
my neck choking me. I have to get home; he has no concept
of how dangerous it can be.

As he starts the car, I remember my kind friend touching
my shoulder as I left, whispering quietly, 'There's activity
there.'

I know I've been put through this VEP test again as *Shaggy
Man* explained that it will tell him if my nerve pathways are
functioning properly. The test is simply an 'evoked potential' –
in other words, an induced response triggered by a visual
stimulus. The VEP tests were the first time I experienced the
sensation of electrodes attached to my scalp. Even in my
emotional state, knowing that they were transmitting my
brain activity roused a kind of morbid fascination in me. That
knowledge, that access to my own inner blueprint, was to
stay with me.

Not surprisingly, the results we get a few days later state
'abnormal activity' for my right eye, but amazingly my left
eye is considered to be 'within normal ranges'. The electrical
currents running inside my head are firing. The messages are

---

* I had a variety of tests when I was in the hospital, but VEPs show
visual function of the optic pathways more efficiently than scanning
techniques such as magnetic resonance imaging (fMRI), so in cases
of optic neuritis, like mine, t[hey are commonly used to assess a
patient's brain activity. VEPs measure the time it takes visual stimulus
to travel from my retina (via my optic nerves) to my occipital
cortex – the occipital cortex being the vision centre of the brain
involved in receiving and interpreting visual signals. They aren't the
diagnostic be all and end all, but the scrolling wave forms VEPs
provide can give clinicians a pretty good picture of what's going on.

coming, but it would appear that I live within a slow-motion film, as the images are taking a long time to arrive.

Fish finger miracle

Jackie is here, wrapping her warmth tightly around me. It's cold outside and winter is tiptoeing up on us, so we leave the house bundled up in scarves and hats. Just walking lifts my spirits and my natural endorphins are buoyed. I can walk more freely now, the stick becoming a theatrical prop as my strength slowly returns.

Back home Jackie makes tea in my kitchen. Lots of people make me tea these days. I also make myself scarce before mealtimes when the kitchen can be a hot and intimidating place. Eerie blue lights flicker under pans as they bubble ominously on the stove. It feels strange, as the kitchen is my natural domain, but even boiling the kettle is a huge achievement right now.

I cry easily these days and today is no exception. The cause of my emotion is often something innocuous, some seemingly ordinary thing, evoking raised eyebrows behind my back. But nothing is ordinary in my life now. Today, it is a dog-eared food list taped hastily onto our larder door, harking back to the time when my house was taken over by kind souls.

I have fingered this piece of paper many times, stroking the corners out gently, refusing to peel it off. In my mind this list holds a magic spell, a list of ingredients that will provide some kind of answer. I won't remove it from the door until I can read it.

When I say I have looked at the list a lot of times, I mean *a lot*, probably hundreds of times. I refuse to let anyone else read the words out loud as that would be cheating. I need it to be a genuine stimulation of my own senses; it must be unprompted and there must be no help. Each time I catch the paper fluttering, I press it flat and force my eyes to find the scrawled handwritten words. They are written in biro pen! What a distant delicacy biro has become.

Today, however, hunched over in my usual awkward stance, nose almost touching the page, a faint magic scrawl starts to drift lazily. One capital letter peeks out of the gloom, flirting elusively, impossible to decipher. Tilting my head and letting out a slow breath, I strain impossibly hard to force something to be there, to *see* these unrecognisable squiggles that I know are letters.

F

It's there! My heart starts to thump loudly.

F - r
F-r-e-e-
F-r-e-e-z-e-r

It takes ages, but the letters ebb and flow beautifully, and I'm reminded of the scrolling text in Peter Greenaway's *Prospero's Books*. I read the whole list deliberately, moving my finger down as a guide, spelling out the words like a primary schoolchild.

2 x t-o-m-a-t-o s-o-u-p

It's slow work and I feel clumsy yet exhilarated at the same time. I glory in '*fish fingers*'. This everyday staple has taken on a new meaning, and the very words are marvellous and familiar; I have just glued it back into the pages of my life.

I was in good humour, therefore, to pick up a get-well card that had slipped onto the mat earlier that morning. We have used the many get-well cards I have received as visual tests, but I haven't had any new ones for a while now. They reveal well-meaning messages, as well as providing patterns and colours for me to hunt out. This one, though, is the most bizarre and exasperating card. However, I refuse to let what is, after all, a kind gesture add to my daily frustrations. Sighing, I look down, and turning over the card turn my sense of humour back on. The card is small, really, really small, probably only two and a half inches square. I smile as I realise that there should be a rule book on how to send a get-well card to a partially blind person.

- **Rule 1** Choose a big card (particularly important if your recipient has also lost the use of their fingers).
- **Rule 2** Choose a bold design, with large letters. Avoid delicate script and washed out colours.
- **Rule 3** Write your message in thick marker pen, avoid biro if you can and definitely don't use pencil.

I smile weakly at the card, admitting defeat, but also feeling somewhat sad, as I would have liked to know who has been thinking of me.

Buckets of sand

Forget individual grains of sand. I am transporting buckets of the stuff today. I am starting to take hold of the reins. I am starting to do things again and my will is galvanised.

I am on my hands and knees, my posterior in the air, and I'm eyeing the multicoloured canvas drawers that line one wall of my children's room. I cannot remember the order of the colours, so this is a good test. I study them so closely that I can smell the musty fabric. I know one drawer is purple, one is pink and I am pretty sure another is yellow. At least, those are the colours I fish out of my memory.

Pulling open the drawers, I haul out knickers, socks and vests, happy that I can identify each item of clothing, if not its colour. Sadly, most of my children's underwear is faded and contains only small patches of colour, so identifying it is particularly hard. I separate subtle shades of grey, black and muddy brown, and am tempted by what I think might be pinks, oranges and creams. Miniature pants mock my incompetence. Holding up an unusually patterned pair of socks, I can hear my mind whisper 'orange' inside my ear.

Moving quickly, my fingers pick out the rough texture of jeans, and a T-shirt that I remember is blue. I pull out leggings, a jumper and a fluffy pink handbag for my daughter. The men downstairs don't know how important accessories are to a five-year-old girl on own-clothes day at school. Taking these items downstairs, I proudly show them to my children. 'I've chosen clothes for you today,' I tell them as they stare mutely up at me.

The children dressed, I now move purposely into the kitchen and feel around in the bottom of a drawer until I touch water bottles. I fill these up at the tap and, grasping for the pockets on the sides of their school bags, squeeze them in. Not finished, I fumble around for the pack of bagels I know is lying on the counter top. I am silently thankful that they are pre-cut, as I don't fancy negotiating sharp knives quite yet. Scraping butter onto the bagels, I am amazed momentarily that these small, insignificant acts are so invigorating and cathartic. I realise that I'm managing mostly by feel and memory, accompanied by a confidence that was not here yesterday. If the men are curious about my sudden activity, they don't comment. I'm glad of that because I don't know or care how I am doing this today; I just sense a shift in myself.

Kneeling down I hand the bags to my children, saying, 'Tell everyone you see that Mummy chose your clothes today. Okay?'

They nod seriously, aware of the enormous mountain Mummy has just climbed, and for a brief moment I feel like their mother again.

Christmas 2012

The approaching festivities are likely to put my coping mechanisms to the test. Protected at home I can shield myself from the outside world, and I decide when and if I eat, what I do and whom I see. Going to the in-laws' beautiful yet gloomy farmhouse for Christmas with the whole family present is going to change all of that. Aside from an appointment with my consultant and the third VEP test, I have not yet had to go anywhere I haven't chosen to go.

The drive down the motorway is creepy because cars take on a shimmering lilac hue as they fly past the windshield, merging into a washed out landscape beyond. Silver bullets are coming at me faster than my brain can process, so closing my eyes is the only option.

I have set some new rules for this Christmas, and they involve doing my own thing. I'm not going to be persuaded into participating in social gatherings unless I'm up to it, and I'm to be left alone when I need it. One of my main stipulations is that I will go out for my daily walk with Ed and the children on our own. I shake my head at the gentle suggestions of some group walks, as I know these will only follow the inevitable pattern of chaos in the hallway as children erupt into tantrums while welly socks are found, and adults mutter over which route to take. My walks are different; they are for breathing in the cold air slowly and calmly, for looking around and assessing any change, without judgement. I do not want an audience for this important daily ritual. These moments of peace outside, safe in the big and small hands of my family, are my refuge, my grounding moments. I don't want anyone else trying to be part of that.

Gently untangling headphones, I feel in my back pocket for an unexpected and very generous gift I have received. An iPod Nano dropped through my letterbox just days before we left. Opening the box, Ed read out a message from all the colleagues I had just spent the last year working with. Turning on the device, he explained that it had been loaded up with music and audio books, and even though I cannot see them, the words 'Get well, Vanessa' drift silently across the screen.

In the main, I am treated with caution and reverence, and a soft sympathy trails around behind me. The daylight lamp-on-wheels that the local low-vision centre has lent me follows me like a faithful dog, brightly illuminating the dark, spider-webbed corners that I choose to hide in. The light fascinates the children and entertains them almost as much as the piles of presents they each receive.

With my brother's help and some last-minute online orders, I managed to buy gifts for my family. The customised silver cufflinks proved to be a step too far, however. I couldn't even retrieve the leaflet that was still stuffed in my bag from the Malvern Show, as the memories were just too raw. I couldn't bring myself to revisit the last day of my old life.

At one point during the festivities I darted across the driveway to my father-in-law's art studio, and felt my way up the cold stone steps. At my request he had left a piece of paper attached to an easel, and a selection of charcoal pencils was

Figure 2 Self-portrait. December 2012. Credit: Vanessa Potter. Photography Malcolm Potter

sitting on the table. Fingering the pencils waiting patiently in a line, I was overcome with a sense of the ridiculousness of this idea, this notion that I could draw myself. Shaking my head, I almost left, but something made me stay. The mirror kept on steaming up as I peered into its watery depths trying to recognise the pale, oval shape and dark halo that must be my hair. My glasses were almost visible, heavy, disconnected circles, but much of my face was missing. I drew myself from memory from a vague concept of what I felt my face was, of who I was. I felt ravaged and exposed, and left the studio shaking and emotional. The drawing remained there for years, guarded by my father-in-law, a physical reminder of all that I had lost.

CHAPTER SEVEN

Painting Project 1

As soon as I enter the building I am wrong-footed. Like so many old buildings that have had new life breathed into them by eager architects, Dulwich Art Gallery is no exception. Glass is everywhere. Once upon a time I would have marvelled at the innovative design blending old with new, but now this building, like so many others, is simply a diaphanous nightmare. But, this is the start of a new year and we have slid into 2013 almost without me knowing it. Even though it is bleak and cold outside, I am infected with a new burst of optimism and drive. *I'm going to get better.*

I approach hesitantly, my vigilant hands primed at my sides, ready to spring up at any moment to avoid walking smack bang into a glass door. Sweating slightly, I dawdle, fiddling inside my handbag to allow my eyes time to adjust, to seek out the hidden seams that hold this translucent building together. A couple approach from the inside and to my enormous relief prompt the glass to miraculously slide open. I make it to the watercolour class that I had booked on the phone, and am presented with a view of bent backs and the quiet hum of hoary exchanges. I am seriously out of my comfort zone here, and feel exposed and naked. A musty charity shop smell pervades the air, mingling with paint and the aroma of instant coffee.

As I hover uncertainly by the door (another menacing glass effigy), the scent of stargazer lilies tickles my nostrils. I hear a friendly female voice call my name, and find myself being led gently by the elbow to a table that has been found for me. I am a curiosity, an oddity due to my comparatively youthful age and sparkly newness. There is a palpable air of familiarity and acquaintance here, and I can't help but feel excluded. It doesn't take long, though, before inquisitive noses start to appear around the board at the end of my table, with kind smiles and hellos following behind.

Two opaque vases and a jug of daffodils are my still-life challenge for today, and within minutes my love of watercolour has resurfaced and these objects have my complete and utter attention. Unable to discern the subtle colours of the paints from the tubes alone, I smear them recklessly onto my white palette, their heady pigments revealing a dusty rainbow. There is no thought, no effort, and my brushes move through the act of painting like well-rehearsed dancers. The daffodils are, I believe, a pale delicate yellow, with perhaps darker tinges of orange at the folds. I can see they are yellow, but to me they are one uniform colour with no hint of subtlety. My brain is filling in my visual cracks, painting an amalgamation of all the hundreds – no thousands – of daffodils that are now being referenced in my mental storage cabinet. I am not painting *this* flower, for in truth I can barely discern its shape, let alone its delicate hues, through the impenetrable mist. Yet, here it is appearing like a magician's sleight of hand on the paper in front of me.

Staring down at it I wonder how much of my life is made up of memory, and the intentional and often unintentional knowledge of how I believe things are, or should be. I cannot *see* these daffodils clearly, yet I can paint them. I am simply a robot performing this art. Sitting back and surveying my painting, it dawns on me that this has been an unconscious act of deception. My absorption was such that I was not aware at the time of how much I was relying on memory. My recall was fast and efficient, tapping a well-rehearsed skill to create this piece of artistic dishonesty.

Kind hands touch my shoulder and mumbled words of praise filter through my thoughts. I cannot accept their generous, yet misguided words because I know that I have failed. The teacher pads over silently, my frown drawing her attention.

'I see you can paint. You have a lovely fluid style.' But I stop her. She doesn't understand. I don't want to paint from memory. I want to paint what I see now, not what my overly efficient memory has inserted. In truth I somehow want to draw what I cannot see. This is not some casual painting class; I am here to find a way to accurately represent how strange this new world appears to me. This is a crude scientific experiment to force my family and those around me to see what I see.

Staring bleakly up at her, I wonder how on earth I can explain this in words that will not make me sound like a pretentious idiot. She seems to get it, though, and bites her lip in thought.

The following week as I wash my brushes I am drawn to a craft drawer to the side of the sink, and peering closely I see that it is labelled 'ink nibs'. The drawer reveals rows of neatly arranged metal pen nibs, and I instantly know what to do. Scurrying back to my table, I lean over one of my paintings and start a slow, deliberate scrawl. The teacher is by my side in seconds, peering silently over my shoulder.

'Ah,' she breathes, 'you're telling us what you can't see. I like it.'

A little later as I lift my head to stretch my neck, I realise that my absorption has been so absolute that I haven't thought about my sight for more than half an hour. It is the first time in months that it has not pervaded my every waking moment. I understand for the first time that I am getting better.

Post-traumatic knitting

Post-traumatic growth is pretty much what it sounds like; some kind of benefit or positive influence that a traumatic experience has had upon a person's psychological function. This is a buzz phrase that was coined in the mid-Nineties, and one that I was to learn about in relation to myself.

My counsellor Noelle and I spend the fortnightly hour I have been allocated through the Croydon neurological rehabilitation service in a stark, white-walled room with a simple table and two chairs. Even this simple, unchanging space allows me an assessment opportunity. I follow the lines of the walls to spot any new details each time I visit. There is a frustratingly bland, pastel-toned picture on the wall that keeps eluding me.

Staring endlessly (and probably obviously) at Noelle's face, I scan for any additional features – a brighter highlight on her skin or hair, or the coup d'état, her eye colour. Our sessions can vary. Some earlier ones saw me work my way mindlessly through almost an entire box of tissues as the grief of losing my sight flooded out of me. This time, however, I spend our session relaying a long list of what I plan to do next. This involves walking a particular distance, reading a new road sign and another painting project, and I also announced with a flourish that I intend to knit.

At this point I glance across at Noelle's face, as I sense a twitch at her mouth. She leans forwards slowly, a more obvious grin now spreading across her face, and whispers, 'You're doing my job for me.'

I have to admit that the knitting was short-lived and extraordinarily hard. However, that was kind of the point. Even using high-contrast, thick wool and thin needles to differentiate the interwoven strands, it was difficult. I was terrible at knitting at school, so it's hardly surprising that I am still terrible at it now. However, I did manage to produce a long, thin, holey scarf for my son that he point blank refuses to wear.

The concept of post-traumatic growth is one that slightly perplexes me, as it implies that I am now in a better place than I was before I went blind. I think it's important to separate out the physical ramifications of losing my sight (as obviously I am certainly not in a better place), but it's interesting to consider if I might be becoming more productive and more capable psychologically. Bringing this up with Noelle, I ask if I could slightly tweak this term to 'post-traumatic respect', and once more a smile creases her face.

Neutrogena 1

One of the most exasperating things I experience is being asked the question, 'Has there been any change?' Yet it is also the most natural question for anyone to ask me. Neurologists, doctors, and my friends and family ask me this question on what feels like a constant rotation.

These questions of progress never become less irritating; in fact, with time they become more so, particularly as, no, there have not been any big changes in what feels like a long time now. In turn, my ambiguous answers only make me squirm at my own inability to describe my visual progress, and the producer who still lurks within me repeatedly raps my knuckles, telling me that there must be a way to evaluate this.

My life is a huge, panoramic painting, one from which I cannot step back in order to gain perspective. I can't don my artist's beret, cock my head and make my hands into a lens, as there simply is no off button for sight. How do I explain that I sometimes feel there has been a change, a sensory shift; but I cannot tell you what it is?

I might have gone on for months burying my frustration if I had not come up with an idea – well, in truth it kind of found me. All of the objects in my house have secrets, and they test me in ways human beings cannot. A plastic shower-gel container residing in our downstairs bathroom gave me an idea. After I informed everyone in the house not to touch it, let alone move it, it now sits in exactly the same position gaining limescale in the corner of our shower.

I decide that for as long as is necessary this bottle will become my own personal visual test. I will never step inside the shower cubicle so I cannot cheat, and I will only look at the bottle from the same fixed distance every day. My eyes are under instruction to hunt out information, tiny bits of data that might not be immediately obvious, in order to plot my progress. I command my brain to seek out the letters that I know are veiled beneath the clouds, and this time my memory cannot assist me.

I am aware of my memory in a way I have never been before. It is a vault filled with images, sliced up by the finest knife; slivers of life stored and filed away. Sometimes a box is nudged off its shelf and a cascade of pictures streams into my consciousness. At first when my brain is shuffling these misplaced cards, the images appear random and jumbled up, but I wonder if a deeper level of my consciousness is trying to prod me, to point something out. When this happens I immediately find myself questioning what knocked the box off the shelf in the first place.

It's unavoidable, but when these fleeting images dance inside my head I know it is something in my outer world that has stimulated it. It is something I have looked at today, but not seen. I am drawn to bright colours – but not all of them, for they have to be just the right potency and hue. Reds and oranges are the loudest, chuckling and waving at me as I walk down the road. In fact I have one brightly coloured memory circling inside my head right now, and it won't be filed away:

> It's 1988 and, kicking my feet in the dust on my way home one sunny evening, I suddenly stop. I am on a rural track that runs behind our house when I find myself rooted to the spot. Fear prickles my neck as I see a large expanse of red blood splattered among the foliage. I am transfixed, the macabre and violent colour draws me in, and I cannot walk past. My heart thumping loudly, I drop my school bag and tentatively step in among the roots and brambles, a perverse curiosity driving me closer. As I reach out my hand I can feel the evening sun warming my back, and in a flash the blood vanishes. I freeze, confusion engulfing me until I understand that the blood was not blood at all, but a fantastic sunset casting its red spotlight on the Earth.

I never saw a spectacle quite like that again, but it burned an image deep within my adolescent mind that thirty years later has decided to pop up – like toast.

February 2013

We're late. One of the monsters is a glassy-eyed zombie sucked whole into the TV, only her physical shell left spat out and empty on the sofa. My son is doing his woodlouse impression as he lies on his back screaming, arms and legs thrashing wildly in the air. I am on my own this morning and neither of my children can hear me. I breathe slowly, in and out, but it's no good and I give in and my shouts ricochet off the walls. Deriding myself moments later, I wonder if I am in fact managing things.

I began taking the children to school myself a few weeks ago, starting the slow amble down Auckland Road, initially with Ed or Mum in tow. Before that I could only manage fidgeting in the passenger seat outside school, horrified that anyone might see me, and ducking down until Ed pushed the children into the back seat of the car.

Thankfully the cold rain has stopped now, so the only hazards as we amble to school are a few unavoidable splash marks on the back of tights. I shrug inwardly as I look up; it's grey today with a flat, unfriendly light that isn't great for me. Trees merge into the houses they guard, lacking enough luminosity of their own to stand proud. Blanket days like this provide little contrast, which in turn leaves the world a very jumbled up place.

I clutch my son's hand tightly through his tiger's paw as we wait to cross the road. Cars still magically appear out of nowhere – they are missiles travelling too fast for me. The children behave perfectly at roadsides now; it's our unspoken rule. They feel my fear and automatically cooperate, so together we weave across the road like a twelve-limbed octopus.

It's still grey outside when Ed and I are sitting in a local Italian cafe at lunchtime. We have started the habit of walking through Crystal Palace Park, then rewarding ourselves with lunch in a different venue each time. I order pasta and stare out of the window. I spend hours staring out of cafe windows as they offer a uniform view of the outside world, one that I can study in sections. Today I am drawn to a sun logo above a local curry house. This sign, which contains a sunburst image with light rays splayed like spokes of a wheel, is slowly rotating from side to side. Leaning over to Ed, my fork poised in the air, I ask him if he can see the sign, too.

'Huh?' is all he manages, his mouth full of spaghetti bolognaise.

'The sign, over there, the sun. Can you see it?'

He sighs as he turns to look, still not entirely comfortable with my constant questions. 'Yes, what about it?'

'Don't you see it moving?' I ask, curiosity starting to creep over me.

'No,' he replies, cocking his head. 'Do you see it...moving?'

'It's rotating from side to side,' and holding my hands up I mirror the erratic circular motions I can see.

Figure 3 One of the many sunbursts that appear to rotate to me. Credit: Ed Potter

'That's weird,' he says, sitting back in his chair and looking out of the window again. 'It's definitely not moving at all.'

My mind whirrs; I understand I have two eyes that read individual messages, which then combine to create one whole image. So why do I now see movement that is not there? Are my eyes not aligned? Is one seeing more, or at a faster speed than the other? I ponder out loud if this stuttering, spluttering vision is due to some kind of crossed wiring?

When I get home I search out all the sunbursts I can find in the house, and I find quite a few. They all move, particularly those with thin black lines, and in some cases the movement is disturbingly violent and questions start to fly around my mind.

Fog

It is inevitable that I look out of my bedroom window every morning, as my husband is physically incapable of opening the curtains himself, preferring instead to remain in his own secluded gloom. So, it falls to me to open up the day, to expose what light I have at my disposal. On the odd grey day, when the garden is shrouded in a floating carpet of cloud, my skin pricks its unease.

'It's foggy, isn't it?' I ask, spinning around, needing to know.

'Yeah, misty. It's kind of flat and really grey…' Ed's voice falters. He knows I need an accurate and precise evaluation from him. I nod as we have established that it really *is* foggy outside, and not just inside my head.

My body prickles and heat creeps around my neck during this fast-fire exchange, because a part of me can't help but remember *that* morning. Looking back, the car journey to the hospital was like two journeys happening in tandem; a filmic split screen of reality. My half was enveloped in a thin veil of mutating vapour, a floating ghostly mantle that drifted and settled on the streets as we drove by. Ed's view was of the brightly lit day he saw through his side of the windscreen.

As I drift back in time I can still hear my soundless screams punch the inside of the car. My mind whirrs and my internal camera clicks, capturing the scenes, the individual frames of this day, storing them away, secreted in boxes. It's very simple really – these images all go into the box called 'fog'.

Pandora's laptop

If I am going to write, I have to stop dictating and open up my laptop again. However, the very act of touching the dormant machine sparks a deep response. This square of metal, now perched on my lap, was once my hotline to the world. I am opening up an electronic Pandora's box.

As the screen first hums and flickers back to life, nothing is legible. The detail on a computer screen is actually quite fine and particularly hard to register for those with low vision. My first attempt results in the laptop being slammed shut, and my tears hastily brushed off the surface.

That was back in January, but I waited a whole month before now trying again; this time, miraculously, I manage to open up a Word document and find myself staring at a blank page. The main problem I have is that I can't read anything in normal font size, and the lettering is very pale grey. As I slowly increase the size I can eventually see my diary entries start to appear without my nose having to touch the screen.

Feb 11th 2013: It has snowed again, unfortunately. I thought the last of the winter weather was past, but obviously not. I hate the lack of light and the whiteout. It's as though I am enveloped in a cloud on these days...

As the weeks pass I slowly reduce the font size; my writing itself is becoming yet another measure of my returning sight, and each decrease gives me a moment of intense satisfaction.

I walked my daughter to school today so I could look through her special art book. Of course she had to walk on all of the garden walls, so our footprints were one set on the pavement, and one set on top of the walls...

And the letters just kept on getting smaller and smaller.

Talking to lamp posts

When you're blind you are pretty much oblivious to judgement. Raised eyebrows cannot curtail my behaviour, as I cannot be embarrassed by what I cannot see.

Maite, my cousin's wife (who I had so much fun with at the Malvern Show), has been visiting me weekly since I came home and has seen many of my strange antics. As we make our slow amble up Belvedere Road, a formidably steep hill close to my home, I suddenly find myself frozen to the spot. Blues have been flickering back into life over the last few months, but what I now see is not so much a flickering, as an entire firework display. In the middle of my murky brown mist there is an effervescent blue fizzing directly in my line of vision. As I stand transfixed, the shape appears to be that of a large blue recycling bin. Imagine a lit sparkler, light dancing all around, distorting the shape, glints and flashes shimmering, emitting a thousand tiny particles of hot metal. This is how the blue bin assaults my senses, and how it draws me in, begging that I touch it. Releasing my grip on Maite's arm, I reach out my hand, whispering the word 'blue'. As my hand touches the bin itself the fizzing stops abruptly. Calm now, the bin is soothed, tongue lolling, its sparkly tremors now stable and inert – a solid blue colour. What's more, the flat blue shade that I see is the most vibrant colour I have seen to date, the closest thing to what I understand colour to be. As I take a step back the fizzing and sparkling fires up again. Looking around at Maite, who in turn is staring back at me wide-eyed, I wonder how on earth I can explain what has just happened.

Since my sight started to return I have experienced many colours and indeed all manner of objects talking to me on a regular basis, and it's the most natural thing for me to talk back to them. Of course they don't literally *speak*, but I have what I can only describe as a profound relationship with colour. I had never behaved like this before I lost my sight, so I can only assume it is a deeply personal response to my heightened sensory world.

These images inside my head seem magical, mystical and utterly confusing. I want to understand this strange, inexplicable experience, to bring it down to earth somehow. For right or wrong, I believe I am forging this unorthodox relationship with the mundane paraphernalia of my life, in order to find it again.

Painting project 2, March 2013

I have a paintbrush gripped tightly in my hand, but actually I am eyeing a pile of glittery paper, cellophane, blue paint and a large quantity of glue that lies temptingly alongside an enormous canvas that I have dragged into our spare room. After raiding my children's craft boxes seeking shiny sequins and glittery card, I close the door securely behind me. I'm going to bring my blue sparkling bins to life.

The desire to build this image, to have it levitate up from the canvas, is overwhelming. Simply replicating my visual landscape as a one-dimensional watercolour is no longer enough. I want more – I want my family to touch it and immerse themselves in what the textures feel like on their fingertips. I want them to live it, breathe it, to *feel* the confusion. I will only be happy when the scene literally jumps out and jolts their senses in the same way it has mine. I know they will never feel the same as me, but showing them is becoming far more important than simply telling them. I start to feel that these paintings might somehow be the start of another parallel journey – something altogether bigger. In truth I want to drag my family inside my head so they can see for themselves what I have seen.

Before long my blue light shows settle down and I now find I almost miss their volatile personality. I knew I only had a small window in order to paint this visual strangeness, so before long I was tidying away the sparkly card and vacuuming glitter out of the carpet. However, that transitory and bizarre relationship with blue has become a powerful motivator, and I want to know what caused such a sensory response.

Neutrogena 2

I have had to be patient, but my daily examinations have eventually revealed some of the larger words on the bottle. Over weeks, bold black letters slowly sharpened and shapes morphed from a blurred sludge into thinner lines. Eventually a single letter started to emerge, and I felt a rush of excitement. I was sure there was a capital N in the middle of the fog.

At the same time contrast started to evolve, too; the difference between shapes and light was starting to separate, to pull apart. Slowly, slogans and advertising hyperbole miraculously started to appear. White type, invisible to me in March, magically emerged in May, and my inner producer meticulously recorded the lot.

Painting project 3

My curiosity is not satisfied, so I start a third, somewhat frenzied painting project. Ordering wallpaper samples online and rummaging through handmade paper in shops, I start to collect a pile of textures, colours and surfaces that I hope, with the addition of textured spray paint, might somehow explain my obsession with objects. I want to bring to life the Nivea bottle that I obsessed over in hospital. Hours later I know I am getting close when, stepping back, I suddenly find myself experiencing a tidal wave of anxiety. Just looking at the montage in front of me triggers intense and terrifying memories. I hadn't expected that.

My startling and violent artworks provoked another response, too. My son, who isn't yet three, crept into the spare room after I had left and discovered my giant picture. A sunny and affectionate child rarely prone to tantrums, he picked up the pointed end of a paintbrush and punched a ferocious splattering of holes straight through the canvas. Most children might try to add their own brushstrokes, leaving a smiling daub of paint; but my son's response was far more visceral. Gazing down at the angry holes I can feel his distress, and embrace his tiny rebellion. Somehow it is the smallest and youngest member of my family who knew instinctively what I was trying to convey, and who responded in the only way he knew how.

Painting and creating my giant artworks only seems to conjure up more questions. Just the act of recreating something when I have limited sight myself is a challenge, particularly if I want those with normal sight to see it in the way I want them to. Then of course my physiological reaction to my own art has been powerful and intriguing. I am curious to know how memories are triggered, and want to unravel where the rules of seeing have somehow become undone. But for now, I am done with painting.

May 2013

I have dreaded this day, but the six-month appointment with Fred has been and gone, complete with pregnant pauses and lip biting. It has brought things to a head, and in the meantime cost me two fingernails chewed off in outpatients, waiting to see *The Big Man*.

I have known for at least eight weeks, maybe more, that my sight has not changed much. It has slowed down, along with my daily commentary. I no longer wake up and decide if it is a grain-of-sand day – it just never is.

Fred calls out for me by my first name, in direct contrast to the other patients I have seen go in.

'It's the miracle woman,' he smiles as I sit down. Today this is a conflicting label, and I can't help but view it a little ruefully. Minutes after the obligatory 'How are you?' question, Fred visibly slumps when I tell him my recovery has slowed down, and my stomach lurches when I see him scratch his head. Looking around I confirm that there have been small changes since my last visit, but nothing huge. Perhaps the carpet is a little bluer, his face a little clearer – but the improvements are small. He suggests a referral to the specialist NMO team at the John Radcliffe Hospital in Oxford to see what they suggest – but he's not handing out miracles anymore.

I want my sight back; and while I read down to the small letters on the eye chart, I don't have the subtleties of sight, the humanness of *seeing*. I miss laughter lines, the invisible flicker of an emotion, a blush colouring my children's cheeks – I can touch them with my hands, but not my eyes. All of these details are lost to me, and I grieve for them. Until this moment that had always felt temporary, so to consider it permanent is devastating and I am not yet prepared to do that.

As Ed and I sit there so close that our knees touch, I can hear our mantra silently downgraded to *some recovery*. How weighty and solid that feels. Of course, I will be forever thankful for the sight I have back, and for that reason it has been a miracle. But dishwater vision is not enough and I have plans to keep on gaining. I'm not quite sure how, but a referral is all I need. It offers one word only – hope.

Neutrogena 3

Finally, it's June and I hit the magic 12 points on my laptop, and with that my daily diary takes on a life of its own.

After weeks of interminable watching and scrutinising, the Neutrogena bottle has also sprung forth its biggest gift. My empirical test came to an abrupt and unexpected halt today when a real miracle happened. After 180 days of watching my limescaled friend, the translucent top portion of the bottle (which I had assumed was a transparent white), today, like a blushing teenager, revealed hidden depths.

Through the thin film of disturbance that still reverberates across my visual field, a new, delicate flush appeared. The top of the bottle is not entirely white; in fact it turns into a pale cream right at the very tip. My surprised family only heard the muffled shriek behind the shower-room door before I burst out and rushed over to the window. Examining the container in daylight I thrust it into Ed's face.

'It's cream at the top, isn't it?' I gasp.

'Yes,' he smiles. 'Guess the test is over then?'

This dirty old bottle has played a valuable role during my recovery, for it has shown the changes I would have otherwise have missed. A smudge in the top left corner eluded me for weeks on end, but it was a stain that I could not ignore. Over time it gained momentum as it darkened and took on a form. By then I could read 'Neutrogena' in bold, glowing type across the belly of the bottle, and I knew that the bottle itself was tapered. Colour had been sneaking its way back in, too; a watery blue had washed over the lower parts like an ocean slowly lapping up the bottle. However, the tip of the bottle remained elusively oblique, merging into the cream tiles behind it. Until now.

Summer isn't just sprouting new growth in the veggie garden; my brain is making new connections, too. It is trying to remember what it is like to see. That mundane bit of plastic, now empty and neglected, has made my heart race. It has prompted me to dance around in glee and today it even made me cry. When my smudge finally made itself known, the word it revealed felt appropriate and somehow right – new.

This has been nothing less than my own homespun scientific experiment, but the visual test allowed me to answer that frustrating question, 'What can you see?' with absolute accuracy; something that makes my inner producer nod quietly and smile.

July 2013

It's my birthday and here I am rocking back and forth on a plastic seat, anxiety sweeping over me in thick, nauseating waves. The snake is back and it is biting hard.

'Bloody idiot,' Mum keeps repeating, but all I can hear are the words the consultant neuro-opthalmologist at the NMO clinic has just uttered, in a chillingly matter-of-fact way.

'You have permanent damage – the sight you have now is all you will get back.'

He has just slammed a door shut, and as I sit here on the platform waiting for our train home, I am scrabbling around trying to find a way to prise it back open again.

The diploma, September 2013

Last year my car nearly got put up for sale, but I refused to let this happen so we paid for insurance and tax diligently while the weeds grew around its wheels on our driveway. My family saw an enormous waste of money, but I saw a goal.

Enrolling on an executive coaching course with the AoEC (Association of Executive Coaching) is a huge endeavour and requires every scrap of courage I have. The coaching course I had wanted to do before my illness has at last become a reality. I'm interested in coaching as a way of helping others achieve goals and aspirations by means of a carefully constructed form of questioning. It's not therapy, but I've seen it spawn incredible results.

While my intention is obviously to pass the course, my main focus is to function like a normal human being. The deal I make with myself is simply that I have to coexist with my fellow students over a period of five months, without them discovering my visual impairment.

Arriving early at the venue, I make sure I sit as close to the flipchart as possible and try to orientate myself in the room. Red pen is almost impossible to read, so I sidle up to the whiteboard at the end of each session to check the written text. I feel obvious and exposed, but in truth nobody recognises that my diligence is in fact a lack of sight.

Mostly I am fine aside from a couple of slip-ups. My train was delayed one morning and in my panic I got lost en route to the venue. Rushing into a room packed full of bodies, I struggled to locate the one remaining chair. I was frustrated that my tardiness, entirely down to my poor sight, would blacken my record, and felt helpless that I couldn't explain this without giving myself away. Therefore I accepted the glance of annoyance from the tutor with a bowed head, knowing that if I'd chosen to do so I could have transformed it into crushing sympathy with one sentence.

On the whole I am not at all shy with people I know, but I have a secret and at times crippling fear of mingling with strangers. Somehow I lose all social skills in these situations, which seems at odds with my confident facade. I have to get over that fear on this course, as mingling is the only way to identify and root my fellow student's faces into my memory. As I can't ask everyone to wear the same clothes week in, week out, I have to create other identifiers. Things like the tilt of their body, the timbre of their voice or perhaps their height all become useful features. I have to get to know these people in a more corporal way, so I can recognise them again. I no longer have the casual benefit of sight to rely on; I need other mechanisms to know these strangers.

The diploma teaches me more than just useful coaching skills. It teaches me that people accept me for who I am, not what I see. It turns out that my sight has nothing to do with the person I am, was or will be. It only affects how easily I can manoeuvre myself in the world, and it has no impact on my place in it. That was a far more important lesson, and it was one I needed to learn.

One of the hardest parts of recovery is the unreal and somewhat artificial life you are forced to live. There is an absence of anything remotely 'normal' for a period of time. We try to do normal things, but it's forced and feels false somehow. I don't know when the transition occurs, when normality creeps back into a damaged life – yet it does. It's rather like lying awake at night wanting to sleep, but the more you try, the more sleep eludes you. One morning you wake up and realise that life is just normal again – and that moment happened to me somewhere in the middle of that coaching course.

October 2013

A beckoning finger draws me over to the nursery manager.

'We need a chat,' she tells me, her eyes closed for most of the sentence. 'Your son has strategies, but not good ones.'

I react with a stab of amusement, but quickly pull on the appropriate mask of reproach. Our son is wily, sneaky and creative, and I love him for it. However, I listen to how he has been sneaking toys into nursery again, and how life has rules we need to follow. I nod obediently and agree to frisk him more effectively before he comes into school each day. As it was I extracted a car from his sock this morning, so his tactics have obviously taken on a more sophisticated slant. I now find myself wondering if the car in his sock was a diversion. If it was, I certainly fell for it.

While my daughter has taken on the role of my protector, jumping in to assist me at the slightest hint of peril, my son continues to thwart me and abuses the opportunities my sight loss offers him. It is an advantage he can make the most of, and I champion his three-year-old ingenuity and ability to work this out for himself.

When confronted about the car at school pick-up time, my smile carefully straightened into a severe line, my son's answer is simple, 'I *dis'tappeared* it, Mummy.'

'Yes, but where?' I insist. He points to a tiny pocket hidden on the inside arm of his Thomas the Tank Engine jacket. As he stands there grinning up at me I just can't help it and my smile pops open again. It's not just me who has developed new strategies; my son is slowly negotiating his way through life, too, and we're both just making do with whatever means possible.

Plans, however, do not always go the way we want them to, and I have to accept setbacks along with my small successes.

The kitchen is filled with loud screeches and a big fuss is made at breakfast the next day. A cold, half-eaten croissant is thrust up into my face.

'Mummy, it's got mould on it!' shrieks a voice in my ear. My daughter's frantic pointing does indeed direct me to some small marks that I can barely make out. I am indignantly informed that they are green and yucky, and her hands fly onto her hips. My son sits quietly in the background telling me he doesn't care, and that he will eat it anyway. I sigh as my little protector and my little peacemaker watch me keenly to see what I will do next.

Quickly, I bin the croissant and, spinning around, announce that I will make toast. I am being watched over, checked by a six-year-old version of myself. The mould police have taken up residence, and while there is humour floating gently on the air, there is also a deep sadness that I quickly scoop up and tuck into my back pocket. Toast it will be today; toast is safe and it won't catch me out.

St Pancras

I plan all journeys now with an attention to detail than I never used to bother with before losing my sight. I can no longer leave anything to chance because I now know from experience that my precarious sight will only let me down.

St Pancras Station is hell, a noisy, overpowering place. As I surface mole-like from the Tube, I stand very still in the middle of the concourse trying to get my bearings, waiting for my recall to kick in and restore me. It's easy to stand motionless and look lost in a train station, as many others there are doing the same thing. I know there are cafes and shops all around me, but my own grimy smog blots out the details. Blacks are dense and brash, dominating the scene. Lines are everywhere, wall edges, table legs and hoardings merging into a checkerboard of movement. The world becomes a large, impenetrable grid in places like this, one that offers only a fragmented view.

Glass is everywhere; it's an invisible trap and something to be avoided at all cost. I spot a Starbucks sign floating among the criss-cross latticework, so I head slowly in that direction. But, not surprisingly, I cannot actually see the Starbucks shop itself; I can only make out the logo. I walk slowly in that direction, allowing my eyes to scan for a gap in what appears to be a giant glass cage. I know there will be a door somewhere, but the pale grey of the metalwork merges into the translucent green of the glass. This leaves me with no discernible openings, so logic comes into play. I can see feet moving around, so I swerve left aiming for where I think a door would be. I am right, and it magically appears just in time for me to walk through it. This is what life is now, walking Harry Potter style through a continually morphing world that makes no sense until I am in the midst of it.

The next challenge I face is my ability to recall muffins. Glancing down as I order a takeaway tea, I know there is no way I can identify any of the cakes on the glass counter below. Glass somehow distorts the light, disguising whatever it is supposed to be on display. I ask for a skinny blueberry muffin, as I guess there is likely to be one quietly lurking down there.

Back on the concourse I resume my standing and waiting with everyone else, as my train platform has not yet blinked up onto the board. A blonde pony-tailed woman is flapping anxiously nearby, and I pick up her movement in my peripheral vision. A slightly stooped, grey-haired man approaches her, and I sense their conversation rather than hear it. The man bows gently towards her, hands neatly clasped behind his back, and I can tell from the way her arms start to subside that he is helping her. As the man retreats I notice the white rectangle of an ID card sway around his neck. I watch him unobtrusively help another passenger, catching his smile as he turns in my direction.

For ten minutes this man approaches a number of people, some flapping like ducks landing on water, some just quietly scratching their heads. I wonder if he will approach me. As soon as this thought enters my head I know that the answer is no. I am acutely aware of the two guard dogs sitting on my shoulders, which growl if anyone looks at me with pity. My confident veneer sometimes pushes people away, yet of all these lost people in this train station today, I could probably do with the most help. The only problem is, explaining why would take longer than I have.

Bedroom window

I have looked out of my bedroom window what feels like a thousand times since all of this happened. I don't so much as just glance out, as study every minuscule detail, filing away the data. My way of looking is an entirely different process. In fact, it is just that − processing. I actually seek out the things I can't see, the elements that are concealed within the dark shadows in the garden, around the side of the shed and underneath bushes. I know small plants that I can no longer see are there, prowling in the shadows. My eyes naturally flick to these darker places, comparing them against the rooftops, tiles and TV aerials.

Yesterday I looked out of the same window and saw something entirely different; I saw what I could see. Momentarily taken aback, I wondered for a tiny flash of a second if my sight had miraculously returned. *Could I see again? Would I even know if I could?*

I shake my head at these ridiculous questions. My logical inner voice tells me that sight doesn't magically reappear overnight, and particularly not after a year has passed. Optic neuritis recovery is slow.

I had to turn back into the bedroom, and *not* see the subtle tones in the oil painting hanging on the wall that my gran left me. I had to *not* see the navy tights tangled up with the black ones strewn upon the bed.

As Ed sauntered into the room I realised nothing had changed. His face was still a mysterious grey, a dull blur of previously familiar features. I was struck by the irony that my brain had recalibrated so efficiently that it had adjusted what it now considered to be normal.

Not seeing has now become so normal to me that I almost cannot tell the difference. These moments are mind-boggling, to say the least, but when I consider it, they do kind of make sense. The thing I have to do now is to find out for myself why that is.

PART TWO
DIARY OF THE SCIENCE

CHAPTER EIGHT

The Science of My Sight

I have received many a raised eyebrow when I say that I started documenting my journey at the onset of my symptoms, literally from the very first day. I had never recorded anything quite like this before, but the drive to do so was overwhelming. I can't explain why I did this, and I can only attribute it to a storyteller's instinct to sniff out a potential drama – even one that was my own. I had no idea, however, that this obsession for answers was going to take me on the most incredible scientific learning journey.

What is explainable to a point, though, is my reaction to my illness. I needed to take notes. I wanted to keep track of and have a way of evaluating this frightening ghost train I had stepped onto. Even when I was seriously ill I knew that one day in the future I would need to pick apart this experience, make sense of it all and share what I had seen. I was fact finding because I knew I was going to ask questions, and that is a trait I had learned from being a producer.

I have a producer's brain, which is one that tends to be analytical and canny at finding different ways of doing things. In my job I had to absorb a lot of information in a short time, often working with new technology, new clients and a wide variety of briefs. I also had pretty extensive visual recall; after all, remembering detail was part of my job. In my line of work there was always some issue that needed sorting, and that was often left to the producer to manage – so I had learned to be an efficient problem solver.

My job was also to not just see detail, but to respond to it. Brains naturally prune away visual noise, but it's likely that my experience allowed me to see and notice detail that others may not have considered important. This detail *was* important to me because of my pre-trained neural networks. Many of these skills are unconscious. I had consciously learned them, but as a producer I saw detail only because I had learned to see detail.

Walking into a room I will always be the person who notices the decor, the lighting, the position of furniture and what the people in the room are wearing; but I would probably not acknowledge the music playing in the background. Working on a painting one night many years ago, I was so engrossed that I completely lost track of time. It was my bewildered flatmate flinging open the living room door that made me look up at the clock in alarm. However, it was not the late hour that had led to her wild-eyed rage, but that, unbeknown to me, my CD player had been stuck on the same song for the last three hours.

Perhaps all I was really doing during my recovery was approaching my own sight loss in this same way. In short, I was just doing my job, except this time it was to be the biggest production of my life.

Sunbeams

Spring is in the air five months after my illness hit, and as I stand at the top of the steepest road in Crystal Palace late one afternoon I witness a strip of golden sunlight cutting through my haze. As I stand there a strange feeling comes over me, an intense desire to share all of the incredible things I have experienced. *How many adults have seen the world reborn to them one wispy layer at a time? Who has seen blue spit and sparkle like I have?* I want to find out the science of it all and link it together. I start to understand why I am recording every moment, every breath of this bizarre whirlwind journey. I am narrating these events not just for myself or even for my family; but so I can show them to everyone.

Staring at the sunbeam I begin to picture how I could replicate my journey in a far more immersive and large-scale way. I think about all of the post-production technology I know about, and how I could use it to recreate this strange visual world I inhabit. Standing completely still, my mind whirrs as I also realise that I need some kind of personal transformation to take place before any of this can happen. These ideas are big! There has to be a moment when I recognise some kind of shift – when whatever is to happen in the future actually starts. I am the first to admit that I'm still delicate emotionally, so I try to think of a way I can conceal myself behind some kind of screen that will protect me. I want to be anonymous, to simply be one patient letting science and art tell her story.

It doesn't take long to come up with the pseudonym *Patient H69*, by taking the first few digits of my NHS hospital number. I am sure being dubbed 'the mystery patient' nudged my inner inquisitor as it lay quietly watchful behind my unseeing eyes, and it certainly influences this name now. My new identity offers the protective veneer I need, and projects a confidence that I don't always feel.

My daily frustration at not being able to articulate what I see has now become my motivation. The paintings were only the start – I now want to use film and digital art to build a timeline of installations that replicate my experiences. I want the wider public to step inside my mind and relive this visual odyssey with me. I suddenly know that whatever I create needs to have meaning and a heartbeat of its own. I want to bring my experiences to life to pique the curiosity of others to better understand their own brains, in the way this experience is making me ask questions about my own brain and visual system.

The first task I take on is to quantify and chronicle my journey to date. The handwritten diary that my unwilling family agreed to write in the hospital is where I start. The hours of MP3 recordings also offer a rich and detailed account, and in fact the MP3 recorder still accompanies me everywhere around the house, continually providing its own form of visual therapy.

The blog

I have spent the last few weeks working out how to build a basic website and the day has come to launch my blog,

which I call 'Talking to Lamp Posts'. I feel a little wary,
though protected behind my *Patient H69* pen name, when
I start posting my story. I have chosen to mirror exactly
what happened to me in the form of a day-by-day diary
account. Before I know it I've had 1,200 hits, but my
readers are not treating my writing like a blog and are on
my site for hours at a time. I soon figure out that they are
reading chronologically – as they would a book. Personal
insights sometimes need time to percolate; and it eventually
dawns on me that I'd been meaning to write a book
all along.

Other things have started to drop into place, too. An
unexpected benefit of my diploma (gained as a byproduct
from all the mingling) turns out to be information. Avoiding
my own illness, entirely but ambiguously mentioning a
'science exhibition', I have managed to discover that several
of my colleagues have links with neuroscientists. Before I
know it, emails are starting to bounce back and forth trying
to connect me to scientists at the MRC (Medical Research
Council) in Cambridge. Soon I'm asked to send a document
that illustrates all of my science-art installations. The ideas I
have on pieces of paper scattered around the house now need
go into a formal presentation – people are interested.

The curator

It's April 2014 and I find myself waiting nervously in the
cavernous hall at the British Museum, looking around for
the curator I am about to meet. Lizzy Moriarty is head of
international engagement at the museum, and has been
introduced to me through an old work friend. I clutch a folder
brimming with sketches of how my neuroscience exhibition
might work. That beam of sunlight ignited this journey, and
the months I've spent dreaming up a timeline of installations
harnessing interactive art to tell my story are about to get a
professional look-over. My exhibition piggybacks my own
case study and introduces some pretty mind-bending concepts

of neuroscience to the public – concepts I am still figuring out myself. I have never designed an exhibition before, so these bold ideas that now span a twenty-seven page document suddenly feel like madness.

I look up as I hear my name being called and see a tall, dark-haired woman in her mid-forties smiling at me. Lizzy is chatty and approachable, and we hit it off immediately. However, it is only when she hears me describe the replica ambulance I want to create, complete with binaural sound-track and sensory stimulus, and the five-foot-tall hand sculpture with chipped nail varnish, that she goes quiet. As I explain the story behind each concept, and in turn what it might teach the public about themselves, her face starts to change. Her interest becomes palpable when she says, 'This could be one of the most exciting neuroscience exhibition concepts I have ever seen, and most importantly – it's one that comes from the patient herself.'

I float out of the museum two hours later reeling slightly, but beaming, with Lizzy's parting words forming a new mantra inside my head – find out the science.

Asking questions

Now armed with a long list of why, what and how questions, I decide to approach the very clinicians who are treating me – the neurologists themselves. These are the medical doctors who diagnose and treat disorders of the central, peripheral and autonomic nervous systems, and who see patients like me. Except that they don't see that many patients like me.

On the last day of my counselling session with Noelle, she gave me a huge hug goodbye. As she walked me to the door I told her about the exhibition and asked if, being a neuropsychologist, she might help contribute to some of the psychological elements. She was thrilled and we met up shortly after at a busy Marks and Spencer cafe. Without thinking I bombarded her with complex questions about how we perceive and understand our visual world. She answered calmly and efficiently, but what she told me only sparked more questions. With her coffee cup poised in the air, Noelle suggested, rather sagely, that I track down a neuroscientist for whom this was their specialist area. Suddenly I realised what a mammoth task I was embarking upon.

I have been to the John Radcliffe Hospital on several occasions now, and after my first shocking visit have started to build a relationship with the specialist NMO team up there, in particular Kay Day, the team's occupational therapist, and Dr George Tackley, Neuromyelitis Optica Clinical Fellow. It's a friendly group led by consultant neurologist Dr Jacqueline Palace, and Dr Palace was the first person I asked for help in understanding my visual system. Her response proved to be very helpful. As we talked I realised that while I had been working in a highly visual industry, I still didn't know much about the very mechanics that had powered my ability to do my job.

When I became ill I also knew very little about autoimmunity, neurology, neurobiology and neuroscience. This shouldn't be surprising, as I suspect it would be quite unusual for any patients to be experts in their illnesses. Being a patient gives me the most basic entry-level understanding of what has happened to my body. In order to be able to understand, I need to become an expert, not in medicine or neurology, but in myself. To find out the why, what and how of my neurological illness, I need to start asking the questions that have been burning inside me.

Before I can do that I also need to learn a new scientific language if I am to be able to understand the answers. I need to speak the lingo. Like a foreigner in a new country, I flounder for the appropriate words to even form the questions correctly. There is no creative license allowed here. I need to find a way to bridge complex principles of neuroscience into a vernacular that will allow me to start having the conversations, which in turn might lead to some answers.

Books

Very sensibly, Dr Palace directs me towards Oliver Sachs as a good mainstream entry point to neuroscience. His book *The Island of the Colour-blind* proves to be the perfect starting place. I find myself absorbed and transfixed by Dr Sach's unusual cases, not just because of their uniqueness and this opportunity to be a voyeur peering into someone else's brain, but because so much of his writing reminds me of what I have experienced myself. Reading about his patients with prosopagnosia (which is the inability to recognise the faces of people you already know) left me dumbstruck. It wasn't just that the real stories were so incredible; I also somehow knew exactly what he was talking about. I am only beginning to think about the incredible complexities of vision and how we perceive the world, but it's a start and it spurs me on to devour more books on that and similar subjects, and in particular anything to do with anomalies of the visual system.

Susan R. Barry's book *Fixing My Gaze: A Scientist's Journey into Seeing in Three Dimensions*, describes her inspiring experience of gaining stereovision after only ever seeing in two dimensions – with the help of vision therapy. Born in Massachusetts, USA, Barry suffered from strabismus (misaligned eyes), and as a child endured several surgeries to correct her squint. It wasn't until her forties that she underwent two years of intensive vision therapy, which created new neural connections and in turn offered her a new 'pop-up' view of the world. Her experience challenged conventional wisdom that the brain is fixed and unchangeable, that programming during a critical period in childhood is permanent. Sue Barry offered me a poignant and revelatory account of the brain's capacity for change, and for neuroplasticity. Her story is also unique and slightly ironic, given that she is a professor in neurobiology and is in fact highly educated in the field of her condition.

As I read these books, my passion for neuroscience is ignited and I realise with an eerie sense of déjà vu that the flat world I first experienced when my sight returned was in fact two-dimensional vision, and Barry's words and subsequent scientific explanations have started me on a road I now have to follow.

The scientist, the beach and me

Things don't happen for a reason; they happen because a collection of isolated incidents collides to create an inimitable energy. While tragedy did stop me in my tracks for a while, other forces were now at work forging a new path.

My exhibition has piqued the interest of my composer friend Michael Powell, who I meet up with for a coffee. We worked together at a post-production company while I was freelancing before my illness. Michael is a laid-back sound engineer and talented composer, and someone I have always got on with. It was he who compiled the hours of music on the iPod Nano that dropped through my letterbox last Christmas.

Out of the whole exhibition, one installation interests Michael the most. This has the meditation tools I used to manage the effects of trauma at the core of it. I want this installation to translate my experience using meditation into something tangible for others to see and understand. This was one of the most vivid parts of my journey, and the visualised mental sanctuary that I now call the beach not surprisingly gives the installation its name.

However, wondering how to portray (or even explain) what my beach means to me has proven tricky. I want others to come inside my mind and float on the sea for themselves, but that is obviously impossible. I believe that any physical manifestation of this magical place would be wrong, so on the back of my experiences having had VER tests in the hospital, I am looking to another device that records brain activity via electrodes placed on the scalp – an EEG (electrocephalogram).* I hope this might be the vehicle to narrate this inner tale instead.

I want to harness this technology by meditating and mentally visiting my beach, and simultaneously recording my brain activity. I then want to translate that brain activity into something artistically beautiful, in turn showing (and with Michael's help) audibly representing my beach in a truly scientific form. I have known Michael for several years but it was only when we talked about this idea that we discovered a shared spirituality. My little team had got a new member.

Fate must have been listening in on the animated discussion Michael and I had, for a week later my diploma contacts finally came through and I got an email from Tom Manly, a scientist at the MRC. My first phone conversation with him in June is slightly surreal, but hugely engaging – a discussion about penguins and the perceived colour of bananas. I decide instantly that I like talking to neuroscientists, and this first conversation ignites a fire in me. Even if I didn't entirely understand the science behind our talk, I knew I wanted to. Tom suggests I chat to a colleague of his – for he's not only a scientist specialising in human consciousness, but also a musician with an interest in unusual science-art projects.

A week later, with a coffee balanced in the crook of his arm, Tom stops his friend Tristan Bekinschtein in the corridor at the MRC to tell him about a strange, but intriguing request

---

* EEG is a means to record brain activity. Small electrical sensors are attached to the scalp, which pick up the electrical signals produced when brain cells send messages to each other. The electrical current that can be captured by EEG is created during excitation of the dendrites in pyramidal neurons in the cerebral cortex. Because the signal of a single cell is weak and has to pass through many layers, such as skin and skull, before reaching the electrodes, only large populations of active neurons can generate recordable electrical activity.

The weak signals reaching the electrodes placed on the scalp are massively amplified before being displayed on paper or stored electronically. These recordings can be assessed by a doctor to see if they provide any unusual results, which might result in a diagnosis.

he'd had from a member of the public. Tristan, who was in the process of transitioning from the MRC as an independent researcher into the role of lecturer at Cambridge University, raises his eyebrows at this. Hesitant but intrigued he agrees to an email introduction.

My emails to Tristan initially fall on deaf ears as he politely avoids my request to talk on the phone for several weeks. Eventually, however, my gentle persistence pays off, and he offers a call with Michael and me. Googling him beforehand I see that his career began as a biologist, eventually leading to a PhD in neuroscience. His lab specialises in 'the non-classic study of the physiology and cognition of consciousness'. While I'm not entirely sure what that means, it sounds promising.

It's a bright, sunny day in July and, with one eye trained on my children careering around the garden at my parents-in-laws' house, Tristan, Michael and I finally talk. The moment Tristan speaks I know we will get on.

'So...Patient H69,' he drawls in a heavy Argentinean accent, 'why are we talking?'

'I have this crazy idea for an EEG art installation – it's called *The Beach*. I want to translate my brainwaves into art and music...'

'Ah,' he drawls, 'you want to use EEG. But, the brain is not harmonious. It wishes to speak to us, but we do not want to hear the sounds it makes.' Tristan erupts into a high-pitched squeal mimicking brain frequencies, which instantly makes me grin.

Tristan's laconic drawl, gracious and inquiring to begin with, becomes peppered with curiosity towards the end. He is clearly becoming more interested in the scientific potential of my crazy plan as our ideas bounce back and forth.

'This conversation has become more interesting,' he finally announces with a flourish. 'I think it is time, Patient H69, for to you to come to Cambridge so we can meet your brain.'

EEG

The electrodes feel wet and uncomfortable on my head now I am left alone in the tiny room, with only the drone of machinery around me for company. I wonder how many others have sat in this same chair, and why they were having their EEG recordings taken. The day has been a whirlwind of laughing, eating and talking about my project, and I find myself musing about Tristan himself. This vibrant, curly-haired man has eyed me curiously all day, with a smile hovering at the edges of his mouth. He has offered up a view of scientists I had not been expecting. Tristan moved to Cambridge from the bustling city of Buenos Aires with his family in 2008, and it's clear he misses the activity of his home town. Arriving with bread, cheese and ham, he makes sandwiches for us all, and the day has been highly social. I can hear muffled sounds outside the room and remember I need to focus; if they are to record what my beach-wave forms look like, it's about time that I breathe…my….way…. back….there.

CHAPTER NINE

The Art of Seeing

I am still daydreaming when my train arrives at the familiar platforms at Cambridge. I'm here to meet a new member of Tristan's lab, a biomedical vision scientist on a research fellowship called Dr Will Harrison.

As we are introduced I am struck yet again by how all the scientists I keep meeting never look like I expect them to look. Will is in his early thirties, Australian and tall, very tall. He is wearing a casual T-shirt and jeans, and as he sits down he drags his hands through his hair. Will has an air of seriousness about him, and I can tell he is very passionate about his subject. As soon as we start talking I'm relieved to have the MP3 player recording our conversation, as within minutes my pen drops and I am gripped.

'Tristan's told me about your story, and I've read your blog – it all sounds incredible and I have to admit I'm fascinated. But…how I can help you?'

'Yes, it was a pretty unusual experience and that's kind of why I'm here. I want to understand the science – the nuts and bolts – of what was happening to me on the inside. So, no pressure – but I'm hoping you might have some answers for me!'

Will's face breaks into a huge smile and, picking a lump of Blu Tac from Tristan's table, he sits back and starts to mould it into a little man as he speaks.

In order to unpick my visual anomalies, Will and I agree I first need to understand how a healthy visual system works, before things go wrong.

Even though I have now read up a little on the human visual system, I ask Will to take me through the important mechanics of how we see – those precious optic tools I had not appreciated when they worked so beautifully for me before. My friends and family are always asking me how my

eyes are, or what part of my eye was affected. This is of course
a natural question (and one I would have asked, too), but I get
many a raised eyebrow when I tell them that in the main, my
eyes were not the problem. As I am starting to grasp, our eyes
are only the starting point in this journey. In some respects I
think it's as useful to get to grips with not just what the eye
does, but what it *doesn't do*. Our eyes are simply receptacles –
vessels that allow light waves to enter our brains and visual
cortex (where the real art of seeing happens). Of course,
blindness can be caused by problems with the eye itself, but
for me this wasn't the case.

As we get into the detail I am amazed at the complex
structure of the eye, and how our visual processing starts.

As Will explains, what we understand to be our vision first
depends on different parts of our eye all working very cleverly
together. Initially, light passes through the cornea, which
is that transparent, dome-like surface that covers the front of
the eye. The cornea refracts the incoming light, allowing the
iris to control the diameter and size of the pupil – which is
simply the opening that regulates how much light enters the
eye. Most people know that the iris is also the coloured part
of the eye – my irises are blue.

Situated behind the pupil and iris is the lens that bends
these light rays to display a clear image on the retina. Rather
cunningly, this flexible and elastic-like lens can adapt to the
image it is receiving by changing its shape to accommodate a
close or faraway object. As a photographer I was particularly
fascinated to learn of this sophisticated inner lens, far more
powerful and complex than any camera I had ever operated.

The retina, found at the back of the eye, is formed of a
thin, delicate, light-sensitive layer of tissue that contains
photoreceptor cells (neurons), which are ready and waiting to
convert light into electrical impulses. As light suddenly strikes
the retina, these electrical signals are processed and nerve
impulses travel from the retina to the brain via the ever-
important optic nerve. Our optic nerve actually consists of a
bundle of around one million tiny nerve fibres made up of
ganglionic and nerve cells. I always think of this as similar to

a fibre optic cable, as it is this biological superhighway that connects our eyes to our vision centres (via the geniculate nucleus), and it was the part of my body that was most damaged.

As I am starting to understand, we *see* with our brains. It is the vision centres inside the brain that do the complex business of processing the light rays we absorb, light that then becomes vital information and forms our view of the world. However, it is not light alone that allows us to see. What we perceive depends on how that light is converted into electrical signals by our eyes, and then in turn interpreted by a vast network of neurons within our visual cortex. In fact, all of our vision is made up of a lot of microscopic pulses inside our head, which communicate via our neurons. It's hard to imagine, but we have millions of neurons constantly producing this electrical transmission of messages, which is vital for everything we think, feel and do.

When one of these electrical signals (a nerve impulse) reaches the synapse at the end of a neuron (the synapse is basically the gap between neurons), it cannot jump to the next one. Therefore the neuron releases a chemical neuro-transmitter, which then drifts across the gap between the two neurons. On reaching the other side it fits like a key into a lock, mooring onto a tailor-made receptor on the other neuron. This landing process converts the chemical signal back into an electrical nerve impulse so it can continue on its way, and all of this activity happens in just tens of milliseconds.

By the end of our first session Will's Blu Tac model is complete, with a thin coil of hair at the top of its head. I am reminded that my daughter also creates hair in this way; I make a mental note to bring Will some air-drying modelling clay on my next visit. In the meantime, I have much to research for myself.

Pillow talk

Shortly after the impenetrable blackness, a watery grey light leaked into my life. Waking up with my visual system literally spluttering into life, it's not surprising that the first physical object I registered was my NHS pillow; but even that was not as simple as it might sound.

Will and I spend much time talking about my pillow, for it turns out that it holds the key in terms of my visual rebirth. I often describe my sight as being reborn, because that is exactly what it felt like at the time, and perhaps my description is not so far off the mark. As Tristan explains this to me, he is munching on one of the biscuits I always bring to his lab.

'When your brain stopped receiving any light signals at all, we can think of that as your visual system shutting down,' Will tells me.

'Yes, it did shut down, and when it came back my first sight was so primitive and basic, it's quite a challenge to describe. I suppose that's why I use visual analogies so much. In fact, I didn't even use the verb "seeing" for weeks as what I experienced initially was more of a one-dimensional sensory response, rather than the rich experience I know sight to be.'

As I cast my mind back I recall the shimmering scene that opened up to me, yet it did not resemble anything I recognised. I try to explain it to Will.

'Imagine holding up an X-ray image, but instead of seeing it as the focus of your attention, the X-ray itself *is* the entire view. It's as though a ghostly film *became* the windowpane to my world.'

'Wow,' is all Will manages.

'Yes, it totally sidestepped reality. But the thing is – my brain told me it must be true, because through this mist a pillow was starting to float in front of me. Now, this pillow didn't have three-dimensional form – it was completely flat, and even staring at it for ages befuddled my brain. On the one hand, I was euphoric that the lights had come back on, and on the other bewildered by this paranormal experience. It took time to work out that I was in fact seeing a pillow!'

'The visual experience you had was the first time that your prior knowledge and understanding of the world influenced what you perceived,' Will explains.

'Yes, I *knew* I was lying on a pillow because I knew I was in hospital, and I could touch it,' I say, interrupting him.

'Sure, in basic terms you received information via your senses, and your sensory system received these inputs and transmitted the information to your brain. But that probably doesn't explain your subjective experience – the way our perception works is more complicated than that. In fact, it's highly likely that your belief and visual memory of what a pillow *should* look like helped shape what you saw.'

'That makes sense. I had to stare endlessly to recognise this *thing* in front of me...' I falter.

'Yes, so, even though you came into the room completely blind, when your sight started to return you already had some unconscious expectations as to what you would start seeing, which in turn helped you to see it,' Will finishes and sits back, watching me try to absorb all of this information.

This remarkable concept that our extensive knowledge of the world can predetermine what we expect to see momentarily blows me away. Of course it makes perfect sense, but I had never really considered it before.

This helps me start to unravel and make sense not just of the pillow incident, but of all the other incredible sights I have seen. However, in 2012 I didn't have a vision scientist on hand to explain it, so my pillow mirage was mind-boggling to say the least, and I did not understand it at all.

Will continues, 'It's hard to piece together any solid theory after the event, but it's likely that your entire contrast sensitivity function was off balance, and in fact probably only parts of that function were operational.'

Seeing my face, Will quickly adds, 'Contrast sensitivity is something we can talk about in more detail next time. It's pretty crucial for you to know about it.'

'Okay,' I reply. 'So...for now, would I be right in thinking my brain switched off, but then when the lights came back on, a load of bulbs and fuses had blown?' Will laughs at this, but nods his head nonetheless and squashes his model man into the palm of his hand.

Contrasting views

'In order to understand your basic visual system coming back online, it's sensible for you to understand the theory of contrast itself,' Will starts by saying next time we meet, then hesitates as he sees my eyebrows shoot up.

'Sure,' I say. 'Give me the basics, then.'

'In the study of vision, contrast is simply the difference in luminance of hue – in other words, brightness or colour that makes whatever you are focusing on distinguishable from other objects and its surroundings. Our sensitivity to contrast allows us to see objects of various sizes and among a range of backgrounds.'

'Ah, yes. I still struggle to separate individual trees within a densely packed forest, so I suppose that's my contrast sensitivity being rubbish?'

'Exactly. Our contrast sensitivity function quantifies our ability to differentiate between relative areas of light and dark. It's worth knowing that this is not quite the same as our visual acuity, which is simply our ability to see the sharpness of detail. You can have 20/20 visual acuity – in other words pretty much perfect acuity – but you can still have diminished contrast sensitivity.*

'I know that all too well. My acuity is fine now, but contrast is still a problem. Its funny that I always felt that my sight returned in layers.'

'That's exactly right – our vision is made up of many complex layers and it works best when these layers all work together cohesively. Our ability to detect contrast is an important building block of our visual system. If there is low contrast between a dark object in a dark place, then that object will obviously be harder to see. However, this is not

---

* Researchers Campbell and Robson discovered that human visual perception can easily be described in terms of our contrast sensitivity to specific pattern sizes presented in an image (spatial frequencies). This is important, because any visual image is simply the physical addition of many spatial patterns.

necessarily because it is dark; even in a well-lit room, an object with low contrast will be difficult to see. We therefore need an appropriate sensitivity to patterns of contrast, given our environment, so that our visual system correctly separates out the different components of our visual picture, in order to then make sense of the entire scene.'

'And that's why I find negotiating a dark stairwell in a dark restaurant so impossible!' I reply ruefully.

'Yes,' Will smiles, nodding. 'Because our basic visual system is intended to be more sensitive to *changes* in luminance, in other words contrast, rather than luminance alone, we can easily see and understand our constantly changing landscape, regardless of shifts in brightness and light.'

So, it turns out that as human beings we have a range of detail that we can normally see – but as my sight was anything but normal in 2012 it's hard to apply standard principles of contrast sensitivity to me. Therefore on the first morning of my 'second sight', my brain kick-started my visual processing with hugely limited resources.

Mind the spatial frequencies

I described lines a lot to Will. Lines took over my life during those first few months of my burgeoning sight, for they delineated the foreign shapes looming up in front of me, but jiggled and shook so much that I sometimes had to close my eyes to escape them. It's only when Will and I start to pull apart my optic world that I begin to realise how much lines, and in fact edges, play an important part in architecturally configuring our visual world.

I notice Will is wearing a worn grey T-shirt today, and he's had a haircut that makes him look younger. I ask him more about lines.

'Our visual brain learns these things in the early stages of our development, when our brain is trying to find statistical regularities, or in other words, what we think our physical world *should* look like based on the information we have,' Will explains.

'So you could say it's a case of joining the dots or filling in gaps?'

'Yes. Hubel and Weisel were forerunners in vision science; you should read up on them.* Thanks to them it's well understood that the visual neurons in our brains like to respond to things like lines – they're the basic elements of vision, and they are defined by contrast. Your brain had hard learned that if there is a small vertical line, then it is statistically likely that there will be another small line vertically above it, and so on. These small line sections are usually connected in one continuous line, but the neurons in our primary visual cortex – the first cortical sight of visual processing – detect each section separately. Our vision is thus efficient by capitalising on the statistical likelihood that lines are continuous. This is true even when some lines appear to end, and it helps us judge the depths of different objects.'

Will pauses for a moment, then points behind me. 'You can try it out now. Choose a wall or corner of this room, and look at one point on a vertical line running from the floor to the ceiling – this example works especially well if there is something blocking part of the line.'

I turn and scan my eyes around Tristan's cluttered office, and find the last part of a vertical line in the corner above his desk.

'Okay, got it.'

'Well, your brain detects the first part of the line, and by linking the contours can infer that the line continues behind whatever is blocking it, even though you can't actually see it.'

It seems so obvious when I do this with Will sitting opposite me, yet we make cognitive leaps like this countless

---

* I did read up on Hubel and Wiesel, and found out that they won a Nobel prize for characterising the functionality of the primary visual cortex and its development (originally in cats). They found that cells were responsive to patterns of light, among other things.

times a day without giving them a second thought. But this was a skill our brain had to learn once upon a time, and we often have no idea that we possess these unconscious abilities. However, when my visual world had been blown apart it was these assumptions that were to help my brain piece back together a fractured visual world, to help me orientate the shapes and lines that were starting to emerge.

'So, if I hadn't had this embedded prior visual understanding, it would have been pretty hard for me to fit what was basically a large jigsaw puzzle back together?' I ask.

'Absolutely!' Will exclaims. 'We unconsciously construct our world every time we open our eyes and use logic to make sense of the visual data we receive.'

So, even though this incredible computation seems effortless to us, a vast amount of internal processing is involved to construct that end scene. I muse quietly to myself that this might explain the enormous mental exhaustion I often experienced in the hospital.

It's not surprising, then, that I spent hours considering the lines and boundaries demarcating my emerging horizon. However, Will threw in another suggestion, that perhaps much of the time my visual brain was not actually detecting *lines* as I understood them per se.

'The lines you describe are probably more accurately – or at least more scientifically – explained as spatial frequencies. In simple terms you were actually noticing the spaces, and the amount of lines or patterns that made up the image. Your visual system was assessing how often a line repeated in relation to the degree of visual space,' Will starts to explain, but then reaches out for his laptop when he sees the quizzical look I know I must have just given him. He shows me two sets of what I can only describe as lines.

'Spatial frequencies are measured mathematically, and are a good indication of the level of detail present in an image. It's a way of evaluating the patterns of contrast that exist in a visual scene, and identifying them helps us label the world.' Pointing to the images, he continues, 'A scene with fine detail and sharp edges will contain more high spatial frequency information

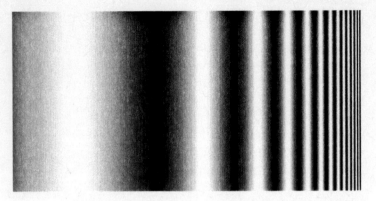

Figure 4 Contrast at Different Spatial Frequencies. From left to right, the spatial frequency of the pattern goes from low to high. [Credit: Will Harrison]

than perhaps one made up of rougher, less-defined visual stimuli.'

Looking at the example, the lines closely packed together have higher spatial frequencies than the ones spaced further apart. You can also see this in the later images I took of Crystal Palace Park.

Looking at more images with Will, an idea starts to form.

'Can *you* create images like this?'

'Yes, actually, using some computer code that imitates biological function that I've written. I use it to replicate certain types of visual deficiency,' Will pauses.

'I've got an idea,' I say, grinning back at him.

Relearning to see

Because, on the whole, brains tend to become less plastic with age, my brain probably had a relatively hard-wired set of neural connections. However, my dramatic sight loss required the systems that learn how to make sense of incoming information to come online again. A complex organic computer, my brain was in fact rebooting itself because, as Will and I have established, this was not the first time I had learned to see. I was relearning how to see by utilising the precise previous knowledge I already had in place.

As a non-scientist it's tempting (particularly given the huge amount of press coverage over the last few years) to view neuroplasticity as some kind of nirvana state. In fact, it's worth understanding that this is simply a normal functioning ability of our brain. The early forerunners in vision science believed that our brains stopped visual development after childhood (the critical period), and that change after a certain age was impossible. Research has proven that to be imprecise, however, and it is heartening to know that our brains can change and adapt, and that is what neuroplasticity is – *change*.

During my visual rebirth, though, what I was experiencing was the most primitive form of vision: low-level vision. It's probably fair to assume that as an adult my brain would have been consolidated, so it may not have been learning anything fundamental in terms of visual statistics – those would have been kept fairly constant across my life and would be relatively hard coded.

I asked Will about infant development as many people have asked me if I initially saw in the same way that a baby does. While it felt somehow comparable to an infant learning to see, there was a unique difference. A baby is unconsciously learning statistical properties – the rules of seeing, like the contour linking we talked about earlier. The whole concept of vision, and how the brain uses visual information, is a skill we learn as we develop. In my case I was learning how those statistical properties now corresponded to this new confusing world. My brain was simply trying to orientate by tapping the old rules of seeing that were lying dormant in my brain.

However frightening this embryonic landscape was, it was going to be fundamental to the regeneration of my visual function. Those wispy images of my hospital room formed the basis of all of my future sight; I would never

have been able to see any new object, landscape or any face
without it.

Stereo

For a short period of time I had virtually no stereovision
(three-dimensional vision). Looking back, I did not articulate
this to those around me in so many words, as the intense and
jarring view that opened up to me was so confusing that to
analyse just one facet of it was beyond me.

Stereovision is simply seeing with both eyes working in
unison, and allows a three-dimensional view. However, any
person with normal vision can mimic monocular vision
(seeing with one eye) by closing one eye for a while, which
then reduces their depth perception. The first part of our
visual system that receives signals is the monocular part of the
brain. Those signals are then combined to create binocular
vision, which is the deep and rich sense of depth that those
with normal vision have.

Stereovision is not present at birth, so again it made me
think about infant development. Trying to focus on my
arms was tricky, as they did not seem to be part of me.
Moving my eyes around in the way I was familiar with did
not provide the same data that it had done before. In babies,
it is not until around five months old that their eyes are
capable of working together to form a three-dimensional
view of the world, and therefore to begin to see in depth.
My ghostly vision offered a strangely flattened world with
no depth, which was both shocking and confusing to a brain
used to seeing in 3D. For some unknown reason, Will
explains that my brain was not integrating the separate
monocular signals at all at this time, which could have been
physical, or a fault at a higher cognitive level. Either way it
proved to be temporary.

It was Sue Barry's first-hand accounts of stereo blindness
that made me realise I had experienced two-dimensional
sight. My descriptions mirrored hers almost entirely, and still
do to a point as I often struggle from time to time with depth.

This is particularly noticeable in the way I take in a scene. Even now, it's an effort to absorb the whole vista, and I find I will compartmentalise and process it as a patchwork quilt. This is why it takes me more time to register things, particularly if I require good depth of field to separate out one object from another.[*]

Street laboratory

As I have described, much of my own vision testing centred on and around Mowbray Road in Crystal Palace. When I couldn't walk very far here, I discovered a smorgasbord of visual stimuli to feast upon.

'Road signs, cars, trees, gates and kerbs all offered wonderfully diverse structures and shapes to look at. Sometimes the cars moved so this altered the test – and became a kind of recognition test! But sometimes they stayed parked for more than a week at a time so I could analyse improvements and changes,' I explain to Will.

'Vision scientists often use stimuli called gratings – you know, the parallel lines I showed you to demonstrate spatial frequencies? Gratings allow us to measure how our visual system responds to very well-controlled visual stimuli. In your case you were not a scientist in a lab, but it would seem that Mowbray Road became your own personal testing ground, and those roads signs and cars all became your own unique gratings. You were being a vision scientist – and you didn't even know it!'

---

[*] In essence our stereovision provides our depth perception, which is how we can judge if an object is near or far away. However, while Sue described the effects of her 2D vision, her visual landscape was normal and included shape, form and colour. Mine, though, was a whimsical replica of what I had once known, and had no recognisable shape, outlines or colour at that time. If our sight is made up of thousands of layers of data and information, my brain was only seeing the first few, tissue-thin layers.

I smile at this thought as Will continues, 'They helped you to measure your reappearing vision, and the attention you paid to testing your own vision perhaps also prompted some of your visual recovery.'

There were also times when I sometimes felt someone was inside my head, with a sharp pencil drawing in details around me. Of course, I didn't have an inner artist, or certainly not quite in the way I felt I had back then. However, over time objects started to gain a new clarity and sharpness as I continuously stared at them. As Will says, it is relatively straightforward for clinicians to record acuity, which is our ability to see this finer sharp detail by using a standard Snellen eye chart that you'll spot in pretty much every high-street optician. But as I have started to understand, my lines were not appearing just because of an increase in my acuity levels; there were other forces at play, too.

Essentially the Snellen test measures our foveal focus. The fovea centralis (which loosely translated means 'pit') is indeed just that – a small, central 'pit' or cavity that is packed with closely knit cones found in the retina towards the back of the eye. I am familiar with the Snellen eye test with its regulated black capital letters descending in size, as I have often squinted and squirmed, trying vainly to read the bottom line.

As Will explains, however, this test does not give the full function of one's vision, and certainly did not tell my whole story. Contrast sensitivity tells us much more; it fills in the picture so to speak. In my case, you might say it was responsible for my inner artistry.

Standing on Mowbray Road in 2012 I had thought it was simply my acuity returning (in other words, that my eyes that were getting better) as my ability to see black and white improved, and more cars emerged.

'This was as much to do with multiple neural brain mechanisms working together, as it was to do with anything else, because your eyes and brain work in tandem at any one time,' Will says.

'So it was naivety that led me to separate them out? At the time I thought it was the actual tones of black and white that were standing out, and that's how I described it to those around me; but it sounds like this probably wasn't the case?' I query.

'I suppose you weren't asking these questions back then! So how could you have known?' Will shrugs, then continues, 'In fact there was an interesting interaction going on between the size of the lines you saw, and the contrast between them. This is pulling on the notion that the visual brain processes spatial frequencies as we discussed before. We can think of a visual scene as being a sequence of overlapping repetitive patterns; in fact, using computer software we can easily decompose any visual scene into any number of patterns consisting of contrast at different spatial frequencies. However, our visual system is not equally sensitive to contrast at all spatial frequencies.' Will pauses, watching me.

'It seems,' he continues, his head cocked to one side, 'that you identified certain patterns, such as shapes, the cars, and lines at lower contrast than other pattern sizes. This just meant that there was some level of detail that you could see when there wasn't much contrast, and of course there were other bits of detail that needed considerably more contrast in order to be visible.'

'So contrast was such a big part of how I saw the world emerging?' I ask, pausing to collect my thoughts.

'Contrast sensitivity certainly helps us see things in the distance, but it is not working in isolation. However, what separates the visible world from the invisible world does depend largely on our contrast sensitivity.' Will sits back and folds his arms as I absorb all of this.

So, catapult back in time to stand with me at the top of Mowbray Road looking down at that ghostly line of parked cars. At this moment in time I am not merely gazing out into the street; in terms of visual semantics, my visual system is becoming more sensitive to information within higher spatial frequencies, and it is these higher frequencies that give detail to a scene. The detail is therefore finally becoming visible to me. But of course teetering at the top of the road I have anything but normal vision, so the fact that I can now see the parked cars due not only to my eyes readjusting, but also to my visual system cleverly relearning how to detect contrast, is pretty amazing. Each day that I stood staring down this street, my contrast sensitivity for high spatial frequencies was increasing. It was stretching its arms, reaching out further and drawing the world back into me.

So, it wasn't a pre-teen sketching inside my head, but instead my own visual system boosting the blacks and whites higher and higher bringing me back towards normal vision. When I lost my sight my visual system didn't respond to these lines, but when it began to recover it started recording higher and higher frequencies, allowing me to see more and more detail in the distance.

Vision scientists like Will are particularly keen on using patterns of lines, although in scientific labs they are known as Gabors – as scientists obviously require a more empirical way of testing spatial frequencies. Gabors, named after their inventor Dennis Gabor CBE, who was a Hungarian–British physicist, are simply a mathematical measure (using gratings again) employed in vision experiments. Back then, though, I did not have access to such specific knowledge or tools, and I had certainly never heard of a grating before. However, knowing this now does go some way to explaining my own unusual fascination, and at times obsession, with seeing and recording *lines*.

Algorithms

Will has something for me. He has altered some photographs I sent him – images that my family took in room three have now gone through his visual programme and depict only a selection of relatively high spatial frequencies. This in turn illustrates a very similar view to the one I had originally opened my eyes to. Looking at them now is eerie.

These images show a wafer-thin X-ray view of my complex visual system starting to reboot. Because it is likely that I was only experiencing high spatial frequencies, there was virtually no light and dark shading to the image, just faint silver lines on a watery palate. This really was a skeletal view of the world – it had delicate bones but no flesh, and did not resemble the world I was expecting to see.

CHAPTER TEN

Synaesthesia

Tom Manly didn't just introduce me to Tristan, but also to Professor Jamie Ward. Listening to my animated descriptions of sparkling blue bins on a Skype call, he suggests that I might have experienced something called synaesthesia. Jamie, a neuroscientist based at the University of Sussex, is considered to be the UK's leading expert on synaesthesia. He has written several books, including *The Frog Who Croaked Blue: Synesthesia and the Mixing of the Senses*, and a number of papers on the subject. Talking to Jamie proves to be a big step forwards in my own personal scientific discovery.

The term cross-modal perception is used to describe the intuitive links and interactions that occur between the senses. Scientists have studied systematic links between music and colour, for instance, and this is where we start to touch the fabulously curious domain of synaesthesia. Several well-known artists, such as Kandinsky and Klee, reportedly described colours and sounds interchangeably as if one sensation inevitably brought on the other. They were describing a condition we now know as synaesthesia. This simply meant that their senses were cross-wired, overlapping each other and provoking a cacophony of visual responses when they heard certain music.

When I describe my visual acrobatics to Jamie in detail, he considers that my strange and temporary experiences with blue as it sparkled and danced was likely to be an acquired form of synaesthesia, as it came on after a traumatic illness and could be accounted for.

Apparently some synaesthetes (whose sensory experiences involve colour) get very emotionally attached to colours. They are fascinated with them; the colour areas of their brains are hyperactive and, like small children, demand attention. This is something that resonated deeply with me. Hearing about and reading up on synaesthesia, I become fascinated. Synaesthetes have hardly any control over the start or indeed the substance of their incredible sensory experiences. I understand this; as if indeed I did experience acquired synaesthesia, it was an uncontrollable yet gripping experience, even if it was transitory. Being given a label to attach to what was one of the most bizarre experiences during my recovery is liberating, and only spurs me on to find out more.

Sensory cross-over

An email pings into my inbox on 9 September 2013:

> *Many thanks for your email and apologies for the delay in responding. As Jamie says, your personal descriptions sound similar to acquired synaesthesia, which is fairly rare and tends to only happen after some forms of neurological trauma or altered sensory input like blindness. I have done research on acquired synaesthesia in the past and am fairly familiar with key papers on the topic...*

Dr Giles Hamilton-Fletcher is a postdoctoral research fellow based at the University of Sussex, studying a wide variety of ways in which our senses can interact. He is responding to an email I sent to the UK Synaesthesia Association, in which I explained my book idea and asked for further help. I'm not surprised to learn that he not only knows Jamie, but also works closely with him; our first Skype call offers some fascinating facts about synaesthesia.

I'm not sure where Giles is sitting, but I can see a black piano behind him and lots of shelves containing books. His blond hair is scraped back into a ponytail, he wears glasses and as I find out very quickly – he is extremely articulate.

'Synaesthetic experiences tend to be elicited, which simply means that they are provoked into being by another associated sense. They tend to be quite vivid and can have strong emotional connotations as well. Each type of synaesthesia works on a pairing system, so for example you could have music-smell or grapheme-colour forms of synaesthesia. In these cases listening to music elicits the impression of strong scents, or grapheme – which are written letters or numbers – and can prompt that synaesthete to see colours in their mind's eye, or even presented over the letter itself!'

'I've read that it's a beautiful phenomenon, and that born synaesthetes see themselves as living in a more vivid and whole world, and can't imagine experiencing anything else,' I butt in.

'Absolutely – most synaesthetes enjoy their synaesthesia and wouldn't want a non-synaesthetic life. The funny thing is, all of us have a subtle form of synaesthesia that operates under our consciousness, within our intuitions.'*

'So if you have a synaesthetic response you may smell a particular scent – like cinnamon, perhaps – because you heard a particular sound?' I ask, remembering my research.

'Yes, but the provocative sense doesn't override the other, music-smell (for example) synaesthetes can hear just fine, and the smell is a complementary experience,' Giles replies.

I find this background information helpful in understanding my own unusual form of acquired synaesthesia. There are many varieties of synaesthesia that are manifested within the brain. In fact, it can occur between any two senses (or more). One of the most common forms is grapheme-colour. Experiencing this, when a synaesthete reads a word they see each letter representing a colour. While there is no common colour alphabet, synaesthetes have their own colour systems, with these pairings between graphemes and colours being mostly constant over time for each individual. So while there is no absolute colour alphabet, Giles tells me that more than 50 per cent of synaesthestes have reported red as being the

---

* In fact, 4.4 per cent of people have synaesthesia, but nearly 100 per cent (including synaesthetes) have unconscious associations between the senses known as 'correspondences' (for example, they know that high pitch equals high locations). So for arguments' sake we all have it.

letter A (it would appear that red is an important colour to many of us). As he states, this is a far higher percentage than could be expected by chance.

Giles continues, 'A lot of synaesthetes who played with those coloured magnetic letters you find on fridge doors as children have grown up keeping those exact alphabet-colour links, in essence solidifying these childhood experiences into their synaesthesia.'

'I *so* get why you want to study this subject!' I reply, and Giles laughs.

'For grapheme-colour synaesthetes, the coloured letters that make up a word can interact in interesting ways – for example, the colour of the first letter tends to tint the rest, so a red "A" in "Apple" creates a red tint for all the remaining coloured letters in that word. Synaesthesia can also aid memory – as in fact synaesthetes can often recall telephone numbers by the combination of colours, which in turn reminds them of the numbers themselves.' He pauses, then adds, 'So, a synaesthete can try to remember a word or telephone number, but it can be on the edge of their consciousness – on the *tip of their tongue*, but they still get the appropriate colours helping them out.'

My smile drops as I hear this last statement as this nudges my own experiences to mind. Here is a distinct correlation between feeling something emotionally before you consciously become aware of it, and I have to know more. I decide to come back to this point later.

Without really knowing their true meaning, I initially described my sparkly light shows as hallucinations, but as Giles picks up, it is more likely that they were 'synaesthetic photisms'.

'These are different from hallucinations in that they are a light or colour response that is triggered by another sense, such as hearing, taste or touch. They tend to be simpler and abstract, and often repeatable given the same circumstances, and are a common hallmark of synaesthetes. This differs from hallucinations, which are more likely to be spontaneous with no obvious external stimulus. Hallucinations can often involve complex things like objects, voices or people, and follow a narrative and are sometimes associated with schizophrenia.'

Considering this, I now see that my blue electrical displays had consistency and that their provocation was actually quite precise, and could be affected once I touched the blue gate or bin. The tiny blue glints of light the gate spat at me were in fact simply an excess of luminosity, but once I touched the gate itself and spoke the word 'blue' out loud, the disturbance subsided and the colour became constant. Giles believes that by using touch I sent signals that dampened down my early visual cortex, which in turn suppressed the sparkling effect. Certainly, this was related to touch, as if I stood back from the gate the sparkling would spring back in all its effervescent glory.

Giles explains, 'In cases of visual deprivation, the normal visual regions of the brain tend to start processing the signals coming in from the other remaining senses. It is likely that the normal visual cortex had to deal with processing touch as well.'

'Hence I got this rather unusual sensory experience as a side effect?' I interject.

'Yes, and it's not surprising that all of this happened directly in front of you, either.' Giles explains that touch can elicit a colour to be seen, and that colour can change according to the touch stimuli (a tactile-visual synaesthesia). Touch-colour synaesthesia is stronger when it occurs in front of us, rather than behind us, so even with my limited vision it still played its part. It's also likely that a similar sensory response was occurring with the coloured cable ties that lit up when I heard the colour spoken out loud, although this used hearing rather than touch as the stimulating sense.

The reality is that I am unlikely to ever know exactly what was occurring inside my brain at that precise moment. Acquired synaesthesia is rare, but it does happen, as I discovered. We all have an innate response to colour, but in some of us our responses are more forceful that in others. As Giles points out, 'Your deeply personal relationship to colour is highly likely to be what led you to experience acquired synaesthesia.'

It's not surprising that synaesthetes compose music or create art – their unique view of the world offers others insights into vivid worlds that they can only imagine.

Seeing red

I describe my early vision as being tinted with a murky brown mist, so colour as I knew it did not really come back into play for quite some time. It teased me, hiding around corners, only to then jump out and assault my senses head on. For months my ability to perceive colour remained hidden away, lurking just out of reach somewhere deep within the subterranean cavities of my brain.

Colour is a manifestation of electromagnetic waves that surround us all the time; it is made up of seven main colours, except that we cannot see them all. They range from violet (at one end of the spectrum with a shorter wavelength), through to red at the other end of the spectrum with a longer wavelength. However, we only see three of these colours, red, blue and green. It is incredible to consider, then, that every shade and hue that our brain-artist magically creates comes from just this limited palate. I think of the hours I have spent in post-production suites tweaking the colours of cars or skies, and the incredible subtleties my brain was processing, all from just a range of three colours.

Of course, all of this processing happens far too fast for us to register it. Light travels at 186,000 miles per second so we cannot intercept, only interpret these visual messages. It actually takes just 10 milliseconds for light to reach our visual cortex and for our brain to start processing that image. Even then we don't see all of the colour spectrum, for some lights such an infrared are not even visible to the human eye.

Our world is colour coded for practical reasons so that we can orientate ourselves and get around safely. Navigation, however, is just a small part of this colour coding. This organic Earth has evolved to create a vast swathe of tones and hues that exist within the plant and animal kingdoms, and provides all living creatures with a visual language. If you think about how you might describe colour you might be surprised. Saying 'red is a letterbox' is not actually describing colour per se, as we only know a letterbox is red because we have identified it as red. However, if you think a little more you might find yourself saying 'heat', 'anger' or 'danger', and then we start to understand why we have colour on our planet.

As I started delving deeper, many things started to make sense. Before my scientific sleuthing I had never stopped to consider what colour really was, as along with neurology and autoimmunity it was not a subject that had ever really come up. This was enough to prompt me to quiz Giles further in order to understand the peculiarities of my recovery. I needed to understand what was happening inside my brain, and to get to grips with colour itself.

Colour has always been fundamental to my job and we frequently used pantone references to match product shades exactly; there is an art to deciphering colour in an empirical way. A technology called telecine (basically the colouring of a film) is one of the most important processes. It sets the tone, mood and feel of a film, but it was often one of the main areas of contention, as we all perceive tones differently.

I decide to do more research on the basics of what colour actually is and how we see it. Colour itself does not really exist – it is simply our brain's sensory response to light. This was a revelation to me, as I had never thought of it in this way before. Colour is simply different wavelengths of light, bouncing off objects within our visual landscape.

As human beings, we understand colour to be a visual property of an object. To give an example, we know that grass is green and the sun is yellow, but the colour we perceive is created by our brain and is not actually an integral property of the object itself. Each leaf on a branch that appears green to us is made up of a particular frequency of light that in turn is made up of photons. These photons are visible to us, hence the leaf appears green, but if the wavelengths were perhaps a bit shorter – then it might actually appear to be blue. Some of these photons bounce off the surface of the leaf, but the leaf itself absorbs the rest. However, it is this process of reflecting light that transmits colour signals to us, and our eyes and brains work in synchronicity to translate these signals into colour.

Context is also important to talk about when it comes to colour. This became particularly relevant during my own personal rediscovery as I was often without the normal environmental cues that a sighted person has. The black corrugated plastic gate with cable ties that Ed and I saw out walking genuinely appeared grey to me. As I did not have the 'full picture' by any means, I could not make assumptions about what I was looking at. In other words, I had to accept the colour I saw without the framework of where it sat.

While I had already had a lifetime of employing the rules of colour constancy (which is the ability to stablise the colour of an object, even across a variety of lighting conditions), I may not have been able to utilise this function fully, as I literally did not have enough vision yet for it to operate as it should. The sun shining on the gate made it 'appear' grey to me, but Ed (with full sight) made unconscious allowances for the sun and eliminated the bleached out areas, allowing him to correctly identify it. I was left in a no man's land of identification, knowing that I should make allowances for the sun, but somehow resisting doing so.

Our ensuing discussions about our own colour perception were probably not surprising, as this is often fundamental to how we perceive our own worlds. However, had we not had that debate (and others like it), I would never have been prompted to investigate the wonder of colour constancy. You could surmise, therefore, that in a peculiar way – we were both right.

I have talked to several neuroscientists about the white/gold dress debacle that dominated Twitter and the media in February 2015, as context famously came into play there, too. It started with a woman posting a picture of a dress on a social media site, and asking what colours others saw it as. This simple question sparked enormous debate, and the ensuing discussions went viral. It brought up emotional and at times violent responses from many people, and illustrated that it is not just Ed and I who disagree over our categorisation of colour. How one describes the dress is based on what you assume the lighting environment to be, and is deeply interwoven into the phenomena of colour constancy.

Of course, we 'name' colours in order to identify them, but this can vary enormously. I have an abundance of pale linens and barley whites in comparison to Ed's one or two creams, and many a paint chart has sparked a passionate debate. So in part, it comes down to vocabulary. However, there is little science pertaining to the physiology of how men perceive colour differently from women, so it would seem that our individual nervous systems, which process colour, are not wired in a universal way. The fact remains – we don't yet have the full picture.

Colour is not just there to make the world a beautiful place (although of course it does). It is there to guide us, sooth us and alert us. It's not surprising that many of us have multifaceted relationships with our perception of colour.

Blinded by the light

I had already learned about our optic hardware from Will, but not in relation to seeing colour, and this is where it starts to get really interesting.

Our retinas contain two types of photoreceptor, which are essentially cells that respond to light. The first are rods, which aid vision in dim and low light and of which we have the most – around 120 million. Then there are three types of cone (red/green/blue) that provide our vision in brighter daylight, which we use for seeing colour and which are less sensitive than their fellow rods.*

---

* We've all heard of cones and rods, right? But cones provide visual acuity and the ability to see fine detail (in addition to colour), while rods allow us to see in lower light levels. At brighter light levels the rods are saturated and can no longer respond.

People who have normal colour vision have all three types of light cone, which is known as trichromacy. However, and this is where the whole concept of seeing colour starts to deconstruct, a percentage have faulty trichromatic vision and therefore have what is commonly known as 'colour blindness'. I have been asked on numerous occasions if I was (or still am) colour blind. The answer is always no, but during certain stages of my recovery I did experience some rather strange responses to colour signals.

Red-green colour deficiencies exist in some people from birth, and this colour blindness is more prevalent in men than women and occurs at genetic levels.* There are around 2.7 million colour-blind people, about 4.5 per cent of the UK population. I have a childhood friend who cannot register the colour green. However, over the years he has been told so many times that grass is green than he now identifies it as green and calls it green himself, even though his brain cannot actually register it.†

I wanted to dissect how colour works in order to understand the enormous emotional impact it has on me. I want to understand how my relationship with colour changed – and how for a while I became colour illiterate.

It's worth establishing how a healthy vision system perceives colour and light, to understand how that changes when it warps into something unrecognisable. It might explain how my senses – completely used to interpreting my world in one way – had to suddenly pull on all their resources to make sense of it in other ways. Colour has meaning to

---

* The difference is due to men having XY chromosomes and women having XX.
† Colour-blind people use all of their three cone types to perceive light colours, but one type of cone isn't quite synced up. In some cases a person has a reduced sensitivity to red light, and this is known as protanomaly. Deuteranomaly is the term for a reduced sensitivity to green light (which, interestingly, is the most common form of colour blindness); finally, tritanomaly, which is far more rare, is a reduced sensitivity to blue light.

everyone, but I was particularly sensitive to it. This might explain why, when colour was taken away, I did so much to claw it back.

Feeling colourful

Ed is sitting on the sofa, listening in. He's pretending not to, but I know he is, and I'm not surprised because what Giles is telling me on this Skype call is riveting.

'Suddenly your brain wasn't receiving the same predictable data that it had been receiving all its life, so it had to change and adapt. In fact, your brain was transmitting just part of the message.'

Giles is starting to explain why I often described sensing rather than seeing a colour, in particular in relation to my daughter's red school dress. 'You experienced an emotional sensation, an acute sensory response – a "feeling" that pinpointed exactly unconscious links you had made before you went blind.'

'So that warmth I felt when I saw red was just half of the message, but it was enough to tap those unconscious links?' My voice falters. 'Refusing to use the word "see" makes complete sense now, doesn't it? I would say that I "felt' there was a colour, but couldn't actually identify it.'

During my recovery this repeated statement received a volley of disbelieving looks, as I suppose it didn't make a huge amount of sense to those silently padding around my house. Except that it does make sense – had my house been inhabited by neuroscientists then surely there would have been a sage nodding of heads behind me instead.

Giles is on a roll now. 'Something called "blind sight" is parallel to what you experienced, so you may want to look it up. It has similarities as it occurs when a blind person, as a result of brain damage in the primary visual cortex as opposed to eye damage, can make accurate guesses about basic shapes and movement presented on a screen in front of them. In experiments blind people have been able to identify and track shapes moving on a screen with a greater accuracy than chance alone would allow. Yet it appears that they do this task despite claiming not to see at all.'

'This reminds me of a moment when I somehow "knew" to pick up a tiny piece of orange pith from the floor back in the early days when I was virtually blind!' I exclaim.

'Sure. It's understood that there are multiple visual routes into the brain. The first is the main visual route that includes visual consciousness – and the act of *seeing*. However, there are also some other minor routes into the brain that can bypass the visual cortex, and can contain this visual motion information.'

Out of the corner of my eye, I can see that Ed has crept closer still.

'So while blind-sight patients would claim not to see an object or its direction of movement, when they guessed the direction of motion using their intuition, they would end up doing much better than expected. This is because the areas of the brain that discriminate visual motion are actually receiving the correct answer through this unconscious minor visual route. So while they can pick out the correct answer, it lacks the context and consciousness typical of vision. It's like knowing the answer to a maths equation is "6" despite never having seen the equation at all.'

Giles's face disappears from the screen for a moment while he hunts out references for me, and I have a minute to ponder what he has said. In my case, I knew my daughter's dress was red, but I had to work in reverse to figure that out based on the emotional response I was getting. My brain was shouting *red*, but in reality it had not recovered sufficiently to transmit that full message, though the signals associating red with warmth (higher cognitive functions) were present.

Giles's face pops back onto the screen. 'There is a lot of research on unconscious emotional links with colour, and this can occur when we experience an emotional resonance with a colour that, amazingly, can still be there even when, like you, a person couldn't consciously identify the colour.'

This has been a highly cathartic scientific revelation for me. It is hard to keep on repeating what you feel so intently and completely, despite the fact that even to your own ears it sounds nonsensical. It was therefore rewarding to find out that my senses were actually complete and wholly correct. In fact, my senses were so efficient that they were transmitting messages to me that correlated to my lifelong emotional relationship with colour.

Green ball

As soon as I related the story (mentioned earlier) of the toy green ball in the green grass, or that green itself often flashed up as red, the colour-blind chant would restart. However, there were other vision principles at play back then, and while I wasn't colour blind my relationship with colour was certainly askew.

To explain my curious encounter with the green ball hidden in the green grass, Giles and I have to delve into the colour-opponent process, which is one of our centuries-old understandings of how we process colours.

He tells me, 'We all have several levels of visual computation going on at any one time. The first stage involves the RGB (red/green/blue) cones and rods, which are sensitive to all light, even low light. These light receptors all feed into the next level of cells, called ganglion cells, and it is here that the messages transmit via our optic superhighway and to the brain. These messages all correspond to the red-green, blue-yellow and black-white colour-opponent theory. This principle is then set for all colour perception within the visual centres.'

'So we have a first hit, if you like – a first response to the light we're receiving – before the next stage of processing happens?' I ask.

'Yes, and our visual system interprets information about colour by processing signals from rods and cones in an opposing way, in other words through the differences between the three receptors. Some colours can be perceived at the same time, but some can't.'

I realise I have stopped writing notes, as this is starting to get rather complex.

Giles continues, 'Red and yellow can be seen simultaneously to create orange, but red and green cannot be seen within one colour as following the opponent-process theory, each ganglion cell can only detect the presence of one of these two colours at a time – quite simply because the two colours counter each other. You'll have seen something that you'd describe as bluish-green, but have you ever seen a single colour that could be described as reddish-green?'

I pause a moment to think about this, but my mind is still a step behind. 'So this means that when I was trying to detect a green ball in the green grass (at a time when I could barely see), the signal containing the red-green message got – what?'

'It got disrupted. However, it is important to remember that your eyes were not damaged, so it wasn't the cells in your eyes that were not functioning.'

I know I experienced very sporadic colour perception during those early weeks of vision, so the colour signals that were making it to my brain were often muddled. I also tended to describe any colour I did see as 'one blanket colour', with no subtle shading, which I mention to Giles.

'Yes, it's unlikely that the grass and ball were exactly the same shade of green, so this instability could well have sparked error messages in your visual system. The ball flashed up as red, which indicated some kind of neural computational mistake. It could be that the colour signals for the ball were stimulating for the "red-green" channel, and that either because your brain saw a different shade of green, or since it was a distinct object, and the brain took this signal, it didn't know how to interpret it and ended up erroneously interpreting this as a red colour nevertheless.' Giles shrugs, peering back at me through my laptop screen.

'I think we can safely say that my visual centres were working really hard that day and my neurons were firing all over the place.' I pause. 'As I tried to process the information the ball flashed up as either red or green until I guess my brain finally settled on green…?'

'It consolidated the information, yes,' Giles confirms, 'but ultimately this is a best-guess scenario. We may never know, and in fact more might come to light in the future as we learn more about our visual system.'

I have learned that colour is simply a human response, in the same way that music is a response to an instrument being played, and we all interpret colours as individually as we do music. Ultimately, there is no music inside a violin; the music is simply your inner ear responding to the vibrating sound waves as they tickle your auditory nerves. I rather liked that concept because while I have an expansive and deep relationship with colour, I am pretty disinterested in music. I guess you can't have both (that is, of course, unless you're a born synaesthete!).

CHAPTER ELEVEN

Memory

Not surprisingly, when I finished my watercolour course in spring 2013 and came home with a pile of detailed paintings, my family was thrilled, regardless of my protests. They insisted that I must have gained much of my sight back in order to create this work; but I know they didn't understand. My paradoxical achievement frustrated me so intensely that I sought a psychological interpretation of what had happened. It made me consider the way in which my memory operated, my knowledge of painting and my preconceived ideas of my still-life subject matter itself.

I have been researching the science of my sight for more than two years now, and I have been privy to my own personal science lessons from an increasing number of highly educated and experienced scientists and clinicians. A Skype call to Barbara Jachs, a friend who is a neuroscience student, allowed me to quiz her on the basics of memory. Barbara gave me some very helpful avenues to research, but the rest of the science is this time mostly my own.

Biting my nails I decide to email what I have written to Tristan, and press 'send' at 11.23 p.m. He calls me at 9.06 a.m. the next morning, and his response to my investigations is positive. After a little help from Tristan and Dr Emiliano Merlo (a neuroscientist specialising in memory based at Cambridge University), the section is done.

The art class

Once upon a time I learned how to paint. In fact, this was (rather embarrassingly) at an adult evening class at the tender age of thirteen. I found myself, a usurper, surrounded by enthusiastic art teachers and retirees who had a collective desire to paint china teacups and dried flowers. For my part I was quietly appalled at their pedestrian choice of subject matter, and sneaked in crushed Coke cans instead.

After time and with practice and much repetition, I became an expert – unconsciously competent in the art of watercolours. It was a talent I could do without thinking, without conscious effort, often staring at my subject matter while my hand moved nimbly over the paper.

However, before that could happen I needed initially to learn how to apply watercolours to paper, how to load a brush with the right amount of water and what colours to mix together to get the tones I required. The repetitive process of learning the rules was an important bedrock and employed procedural memory, which is finely interwoven with the use of my fine motor skills. This basis of learning allowed me to eventually translate this into an implicit and automatic knowledge.

Looking back, I see that I picked up other painting techniques not learned in that evening class. I absorbed better ways of applying the paint, of using light and space within a composition without ever being able to pinpoint where I gained that information. The class gave me the basic techniques, but the fluid expertise and proficiency came later on, after hours of painting by myself at home and visiting countless art galleries. These skills were even influenced by my perception of the world around me, which included my passion for photography, the composition of my surroundings, literature and all the mundane snippets of life that stuck to me like Velcro. The person I became was in part an amalgamation of my own memories. All of the visual trivia that I sucked up and stored led to my interpretation of my aesthetic world. My memories are part of my psyche, and they influenced the language I use and eventually the career path I took.

There are several areas of our brain that are involved in procedural memory. The cerebellum (the part of the brain that is involved with muscle and motor control), the striatum (part of the forebrain concerned with cognition, motor and action planning, decision making, motivation, reinforcement and our reward system) and our limbic system are all involved with supporting procedural memory. By putting myself back into a familiar context of painting, I was in fact firing up my limbic system, as I had previous emotional links attached to painting. I have fond memories of painting flowers, utterly consumed in my bedroom as a teenager, so all of this will have come into play again and would explain why I so enjoyed and got lost in this new painting experience. My limbic system is responsible for my emotional responses, so not surprisingly is deeply ingrained into memory formation

Memories have a kind of three-part process as they are formed, stored, then retrieved, and come in many shapes and guises. So, while I could not fully distinguish the daffodil in front of me, all my previous knowledge of painting daffodils was retrieved when the neural connections that I made years before fired up again. My knowledge of daffodils came about through explicit knowledge as I had got to understand what a daffodil looked like – its colour, texture and size – from exposure throughout my life in a variety of different ways.

I found myself squeezing out colours purely on the basis of past knowledge. Burnt umber provides a great tonal companion to cadmium yellow. I could barely identify these colours, yet I knew to use them.

However, creating a memory is a complex process; it is actually known as memory consolidation, and occurs during the first few hours after learning or encoding a new piece of information. Once I had the basic watercolour skills, I could apply this knowledge and experiment further and find my own artistic style. This neural data was then fixed into my memory, and became something I was unlikely to need to learn again as it was forming part of my long-term memory.

But I can't talk about memory without mentioning the mighty hippocampus (this is the part of the brain responsible for emotion and memory). It is widely understood that many of our memories are formed in this part of the brain, and it is where the process of memory consolidation takes place.

My art class was an example of how a particular skill, when repeated over and over again, can become deeply embedded into the mind. Following on principles of neuroplasticity, neurons firing together create stronger connections, and the more I painted the more those connections were reinforced and became permanently responsive to each other. The signals were stored away and could be retrieved in the future – and as in my case, when sitting in an art class twenty-eight years later.

It was this permanent (yet automatic) wiring that I could not consciously override. Those deeply rooted pathways would not allow me to truly paint what I saw, and I ended up painting what I knew was there instead. My brain made assumptions as to what this daffodil (and indeed any daffodil) should look like, based on my past perception of this friendly yellow flower (the very fact that I describe this flower as friendly tells me so much!).

While the intricate curls of the daffodil's trumpet were too subtle for me to discern, I could paint them from memory. I found it almost impossible *not* to fill in the flaws in my vision because my brain had so much information to access.

This experience also introduces themes of false memory, something Tristan was quick to point out when we talked. When a memory is retrieved it is reconsolidated; which means that it is rewritten. Thus an eyewitness in a court case is not as reliable as we might think, as our memories can change. False memory is when we recall a memory that in fact did not happen. This is a very human trait, and isn't intentional, but often when we're questioned after an event, we change our recall of it. If you think of the story that you always tell, consider how it might have changed over the years of telling – probably due in part to the responses of others. These suggestions can sometimes become embedded into the story itself over time. False memory is linked into our expectations, again bringing in the idea that your previous knowledge and expectations can influence what you see, and indeed what you thought you saw in the past.

In my case all of these complex systems were only made apparent because I was conducting my own mini experiment. This meant that I had creative objectivity, as I was consciously aware of what my brain was doing. So, you could say there were several obstructions to the success of my experiment; not only my inadvertent and automatic extraction of my deeply embedded watercolour skills, but also my previous assumption of what I understood a daffodil to look like.

It turned out that my past was always going to influence my future, and it would be impossible for me to paint what I could not see as my brain would just not allow it. It was a battle I could never have won, and in fact my scrawled notes at the bottom of each painting were simply my way of conceding to the power of my too clever brain.

CHAPTER TWELVE

Making No Sense

I met Nikki Adams initially in 2014 in York. She is considered to be a leading expert in the field of neurophysiotherapy, and in fact was recommended to me by several other clinicians also based in Yorkshire. Along with Kay Day, the softly spoken NMO occupational therapist I met at the John Radcliffe Hospital in Oxford, she has helped me understand the strange physical sensations I experienced during my illness and recovery.

Nikki, a slim woman with an elfin haircut, talks about the brain with enormous passion. Our first chat spawned a raft of emails, and she has always been intrigued by how I responded to my illness.

'I have seen other patients with similar symptoms and problems to you, but few retained the cognition, memory and insight to report and describe them with the detail you have,' Nikki explains.

In the same way that I got to grips with our visual system I now go back to basics, initially with help from Kay, which we do via email correspondences. Kay's positive and gentle manner towards me as a patient made her very approachable when I started to ask more in-depth questions. She explains that as busy human beings, we're not always consciously aware of what our body is doing at any one time. Yet our body's movement, that intricate pulley system that makes us walk, pick up objects or run up stairs, is only operating because of the electrical signals transmitting to and from our brains. If for some reason those signals get interrupted, or even partially interrupted, then all manner of strange and unexpected things start to occur (as I found out). It was probably because I only had partial interruption and had some signals still getting through that my confusion, and indeed my curiosity, were piqued in the first place. As Nikki points out to me, had I had no sensory signals, there would have been nothing to work with. It was the strangeness – the baffling signals I experienced – that became my motivation to understand what was happening not only to my visual system, but also to my body.

It is commonly believed that we have five main senses, but we also have additional, sensory receptors that transmit information to our brains, such as our proprioceptors (tiny sensors that detect position and movement), nociceptors (cause pain sensations) and equilibrioceptors (help to regulate balance). This concept makes more sense to me now, as when my vision was no longer my primary sensory channel other sensory organs that I was less acquainted with came to the fore, and became new eyes.

'I'd say in particular your proprioceptors got a pretty good workout,' Nikki starts off saying. 'These sensors are located within your muscles, joints and inner ear, and feed back environmental data to your central nervous system, but they may also have been affected.'

'Yes, I know it's my proprioceptors that kick in when I close my eyes in a yoga class,' I joke in return.

'Absolutely, they keep you upright. What's more, this information doesn't get transmitted singularly, but is integrated into all of your sensory input, to create a full picture of where you are physically at any one time.'

Kay had already explained our ability to walk – and that this requires more than just our muscles, bones and joints to work, as in fact our brain must instruct our body to move. However, as I find out, this is not a one-way street; there is a continuous transmission of signals flowing from the limbs to the spinal cord, brainstem and brain itself. So, in order to walk normally and purposefully, this process is controlled at a number of centres within our nervous system. Purposeful movement, in other words movement requiring conscious effort, is generated in the cerebral cortex. However, automatic movement is generated at the spinal cord level, and to some extent is learned (although some automatic movements are reflexes). A professional tennis player once learned how to swing his racquet, but after a while he became unconsciously proficient in this act and it became an automatic skill. As you walk down the road chatting on the phone, your act of walking is automatic.

Nikki expands further on what happens on the inside.

'If there is damage – in your case inflammation causing transverse myelitis at the level of the spinal column or brainstem (where the spinal cord connects to the brain) – then movement becomes impeded. Both automatic and purposeful movement are modified with input from the basal ganglia and the cerebellum (the centres of the brain concerned with movement). These adjustments allow us to have normal and coordinated movement. Our movement is also affected by the sensory inputs we receive, and all sensory signals go to the thalamus (which acts like a large telephone switchboard). The thalamus does several jobs, but is responsible for our sensory and motor signals, along with regulating our levels of consciousness. So, in effect we move according to how we feel. If our legs feel freezing cold, then our movement is likely to be clumsy and awkward.'

'Actually, awkward is a good word,' I interject, 'as at times I felt as though my body belonged to someone else – it was unnatural and cumbersome and it didn't respond in the way I was telling it to.'

In my case I had to think much harder in order to make what used to be automatic movements become natural again. I actively retrod this process in order to perfect my basic motor movements – simple things like holding a fork, or swinging my foot forwards in order to stride. In a similar way to my vision, my nervous system had to fire up old synapses and relearn what it already knew.

Of course, I not only had damage to the nerve pathways in my spinal column, but also to my vision centres, so the normal visual data that would help me balance and orientate myself was missing, too. My vestibular system, located inside my ears, provided information about where my head was in space, and helped me adjust my head, eyes and body in order to stay upright (which was a challenge early on). These systems were hard at work trying to compensate for the absence of co-workers. The expected signals were just not getting through – there was a neural roadblock in place.

'I'm curious to know why I knew to keep moving all the time,' I say to Nikki on one occasion.

'I think what separates you from other patients I've seen is that most don't try to find out the solutions for themselves,' Nikki suggests.

'There was no way I was staying immobile!' I exclaim. 'Though having said that, my constant shuffling and physical testing made my family worry that I was overdoing it. But in my mind – I was *redoing* it. I knew I was relearning how to do all the things I once used to do without thinking, and I wanted it back quickly. For some reason I felt as though I had a huge clock ticking next to me.'

As Kay tells me, my physical symptoms are best described as sensory loss – a loss of sensation in my fingers, hands, feet and parts of my legs. I had patches of numbness higher up on my thighs so pronounced that at times I couldn't even feel a pinprick. I also had minor motor loss in my right foot, so that it dragged along the floor, as I just couldn't seem to flex it to make it stay in the right position to walk normally.

My sensory loss was a source of huge puzzlement. I experienced very specific things that I discuss with Kay and Nikki on separate occasions. I describe my fingers feeling as though they had rubber tips on them and had been bound up tightly with Gaffer tape. In comparison, my feet felt as if they were encased in blocks of solid ice so restrictive and hard that I couldn't even wiggle my toes.

Both Kay and Nikki agreed that the intricate wiring that conveys messages to my brain from my spinal cord was getting confused. As Kay very articulately put it, 'The normal chatter that travels along this tiny cabling crossed over, and other conversations got interrupted. Messages asking *How does this feel*, or *Is this cold or hot?* were just not getting through.'

It was rather as if my temperature gauge had gone AWOL, and my brain was guessing the answers and getting it wrong.

But this is just the start of our sensory system; information about what we're touching is encoded before it even reaches a nerve cell. This happens in the thousands of special sensors hidden in the skin. We have lots of different sensors – some detect heat and pressure, while others detect damage – and these tiny informers are located at different points on the surface of the body. Each sensor has its own superhighway that travels on a unique route via the spinal cord and brainstem to specific locations in the brain itself, where it is interpreted. I had never been aware of these nerve centres before, but I was to become very well acquainted with them as I tried desperately to return normal sensations to my limbs.

After a relatively short time my attempts at tiny movements were successful. Even a tiny twitch of one toe brought shouts of glee in my little room. The parts of my leaden body that would move were forced into even harder Pilates positions. I knew if I wanted to walk normally again, I had to start teaching my body what movement was.

I also tried to stimulate my feet myself. This was a casual suggestion made by the hospital occupational therapist as she left one day, but my family and I took this homework very seriously indeed. All manner of stimuli was brought into the hospital, and had Matron spotted it I'm sure it would have been confiscated as contraband. Cotton wool, a spiky garden tool, a beaded statue and a kitchen scouring pad were all rubbed methodically up and down my feet. At times this was excruciating and tantamount to being tied down and tickled, as the sensations that did make it to my brain were so extreme and uncomfortable.

However, over time I started to separate out the soft objects from the hard ones, and my toes began to move. It literally felt as if the ice was breaking up, and the hard cement changed to feeling like gritty sand. The feelings were incredibly authentic, and I am still to this day stunned that my brain could recreate such intensely real sensations, which were, of course, entirely imaginary.

Nikki tells me, 'Your brain was most probably referencing old sensory experiences in an attempt to understand what was happening. As more sensory elements came back online, your brain translated these into the grit and sand you felt.' Absorbing this information, I find myself musing that this mirrors what had happened with my vision, too.

I often found myself talking directly to my feet. I think this was the first time I adopted this habit, but I am convinced that it helped. I would tell my feet when something soft was rubbed on them, or when something was 'scratchy' – somehow reconnecting the knowledge I had to the feeling I was experiencing. I think because I could not see or feel my own feet, talking to them was the next best way to connect again. I called this 'joining the sensory dots', and it's a term I have come to use a lot.

Nikki immediately comments, 'That's a great way to describe it – it's likely that this did have some effect, as this technique is used post-stroke to re-educate the sensory system. Plus we know that our vision is richly connected to language, so this was all part of you stimulating recovery.'

I also knew that our unconscious mind hears every word we say, and that in part forms our belief system, so I suppose I was utilising this knowledge. I have no way of knowing if the verbal barrage made any measurable difference, but letting off steam made me feel better, so in terms of my mental well-being and biochemistry, that was a good thing.

When it came to my hands, we tried different things. My friends massaged my palms, rubbed handcream into them and painted my nails. All of these tacit activities were performed to provoke those tiny sensors on my skin to respond. As Nikki explains, it was the repetition of these acts, and the variety, along with focusing my attention on what I could feel, which enabled me to feel more normal responses again. I have never been a passive personality, so it was natural for me to want to touch those around me, particularly considering my sight loss, and these hand manipulations became second nature to us all.

Nikki points out that this is a key point, 'When a human being experiences loss of one sense they (literally) grab onto whatever other sensory stimulus is available to them, and in turn these senses are heightened and enhanced.'

The nail painting was another matter entirely. The sensation of Marissa painting wet nail varnish onto my nails was extremely odd. However, we were simply trying to stimulate my visual response to colour. Looking at each nail, each painted differently, I would parrot the colours out loud. *Blue* I would tell my thumbnail; *purple* I would whisper to my little finger. This was in no way feminine grooming; it was a serious experiment.

The speed of my physical recovery was down to a number of factors. I had a good level of care in the hospital and had a reasonable level of self-awareness, both physically and mentally, combined with a personality that liked to find solutions. I also had the benefit of a mother who was a retired physiotherapist, and from whom I had absorbed secondhand knowledge for most of my life. I knew that muscles atrophy in no time at all, so it was this that motivated me the most to start miniscule movements while lying on my bed. It was extraordinarily hard to even raise my elephant-like limbs an inch the first time I tried, but even trying rewarded me with more data. When my raised leg caused me to wobble, I had to use my hands and the muscles in the back of my legs to steady myself, and even that was progress.

I was lucky that my mother had spent her career helping children recover and improve mobility, so I definitely had some advantages. We were both out of our depth at times, but she encouraged me to keep moving, and she knew that what you do in the early days can set the scene for any future recovery. I know I am lucky, as not all patients regain the full mobility I did. Chasing our children around the garden now is something I no longer take for granted.

I suppose looking back I wonder if perhaps the very human instinct to bat away what we don't understand should be reined in when dealing with complicated neurological symptoms. Sometimes it felt as if the logical minds around me didn't listen to what my body was desperately trying to shout. At times I must have sounded like a patient who made no sense, yet now I know it all makes complete sense – well, in as much as my investigations have scratched the surface of our highly complex sensory system. My body was in disarray, but had I known that the bizarre sensations I was feeling were explainable – then maybe I wouldn't have suffered so much mental unease. But then again, maybe I would never have been prompted to go off and find it all out for myself.

CHAPTER THIRTEEN

Cambridge Science Festival 2015

I can't decide if I believe in fate or not, but I can't deny that there were other forces at work that propelled individual and unrelated people who would never have even met, let alone come together, to create the impossible. Three women were at the core of it all.

Barbara Jachs, a young Austrian neuroscientist, first met Tristan when he gave a talk at her university during her masters degree. Barbara had discovered mindfulness meditation as a means to cope following a rocky patch in her personal life and, not surprisingly given her field, this sparked a deep fascination for the effects of mindfulness on the brain. It was something Barbara was keen to pursue in Tristan's lab, though his initial response to this was somewhat sceptical. However, unbeknown to Barbara and I, there would be one final woman to sit in Tristan's office passionately talking about meditation.

Dr Ann-Marie Golden is a psychologist (and an old colleague of Tristan's) who specialises in both teaching and researching mindfulness. She had just returned from a stint in Canada where she had completed a PhD in mindfulness research. She too drank Argentinean coffee with Tristan in the autumn of 2014, and this time Tristan's frown turned into a smile. It took three women each with different approaches, but all with a collective passion for meditation, to make the cogs finally fall into place, and with that *The Beach* project was ignited.

The gallery

My team has grown even further. Jackie is not only my friend but also one of the most resourceful producers I know. She and I are on the phone daily talking about blackout blinds, signage and practicalities of the art gallery the Cambridge Science Festival organisers have given us to host my EEG installation. Lizzy and I are talking regularly, too. With her experience and contacts via the British Museum, she has helped me coordinate the structure and flow of the exhibition. We have recruited a science writer and an independent evaluator who both specialise in science communication to help us make the most of this opportunity, and the neuro-science students in Tristan's lab are being recruited to man the exhibition itself.

It didn't take long for Tristan to introduce me to Barbara and Ann-Marie. Barbara and I are soon busy creating a website, PR materials, scientific questionnaires and all the necessary paperwork required to put on such a complex and technical project. We are running *The Beach* (the name of the installation) as a pilot version – to see if we can even get the concept to work. We plan to hook up four members of the public in separate hourly sessions, aiming to channel around twenty-four people a day over the five days we have the gallery for. We're planning a blacked out sensory controlled area where our participants can watch (with the help of headphones) an animated film that explains my story, and shows the brainwaves we recorded while I meditated on *The Beach* translated into moving art and music. Then, while they are hooked up to our EEG machines, we will invite them to try out mindfulness meditation for themselves for a five-minute period while we record their brain data. After we unhook them from the EEG machines, we'll show them their own brainwaves instantly translated into music and art on screens, with our scientist students on hand to answer any questions.

We are also going to collect scientific data from the participants, allowing Tristan and Barbara access to some initial, tentative data on how just five minutes of mindfulness practice might affect the human brain, and whether we can intentionally alter our state of mind using mindfulness. A key component to this is to create a comparison for our participants. In order for them to know they achieved a meditative state we are going to take a baseline for one minute where they just rest with their eyes closed, so they can see for themselves that resting and meditation are not the same thing on a brain level.

As this is a science-art communication project, I decide to involve my children's school in creating some sculptures that might help our audience to start to identify our colour-coded system for the brain frequencies they represent. Barbara and I spend an exhausting day running a hugely popular neuroscience-art workshop with children from years five and six, which results in a spectacular set of head sculptures and drawings that now decorate our gallery entrance.

The brainwaves

Tristan and his team spent several long nights condensing the data received from my EEG session into the four main neural bandwidths – beta, alpha, theta and delta. These bandwidths, also known as brainwaves, collectively represent our state of mind and offer a good indication of whether I was in a meditative state or not during my own EEG session.

All of our brainwaves are active at any one time, so Tristan and his team were looking for those that were dominant, whether beta waves, which would indicate that I was more active and mentally alert, or theta and delta waves, indicating that I was deeply relaxed, calm, meditative or even quite drowsy.* You could say these brainwaves provide a map for

---

* Although each individual's brainwave patterns are unique and complex, some patterns can be generalised and associated with certain states. We interpreted the different brainwaves as indicators of a state of consciousness. The waves can vary in their speed and amplitude. The fastest (or highest frequency) waves measured in the exhibition were beta waves, the slowest delta waves. Originally, these waves were thought to be epiphenomena – they were not thought to possess any function themselves but to occur as a side effect of neurons communicating. In recent years, however, this view has been revised and it has become plain that the oscillating property of neurons has an important functional role.

Tristan to find my beach. This code then provides the blueprint we need – a way of measuring the state of mind of each member of the public who we are hoping will participate in our project.

The art

Dan Shorten is in his thirties, wearing a woolly beanie hat and sloppy jeans when I meet him at the Guildhall School of Music and Drama inside the Barbican Centre in London. He is a lecturer in the Technical Theatre Department, but came across my venture through a friend. One of the biggest challenges I have had is finding the right expertise to translate the EEG data we have collected from my brain into moving art. This is a step away from the post-production I know about – we are venturing into the mystical world of digital interactive art that requires complex programming and coding. Dan scratches his chin absentmindedly as we chat, and I can't help but notice music students humming and tap-dancing their way past us in the vast corridors. It doesn't require too much persuasion and I soon see him smile, for *The Beach* is the perfect vocational project for his students.

Before long Dan and his team are creating the colour-coded animation we have collectively designed to represent each bandwidth, and he's working with Tristan to modulate this animation according to the EEG data that will power it. No formal language exists to interpret and in turn translate the undulating waveforms outputted from an EEG machine into art. EEG isn't designed to be beautiful – it is designed to be functional, but we're planning to change all of that.

The music

Michael, my composer friend, has been putting his music skills to work, and has provided a set of audio sound beds to help translate my EEG brainwaves into music. These sound beds are essentially a set of synthesised notes that each represents in an auditory fashion the same EEG data Dan is translating into art. They are in effect an audio alphabet. Synchronising the sounds as closely as possible to the evoked data creates an individual composition as unique to each person as, well, their brainwaves.

Unable to use colour coding this time in order to help our audience identify which neural bandwidth is dominant, each notation reflects either the high fast activity of beta down to the slower methodical sounds for delta. Creating the notes at the same hertz (Hz) frequency as the brainwaves themselves also helps translate their meaning. Hertz is a measure of cycles per second and is used is in relation to measuring electromagnetic waves (brainwaves), and also in sound production. To give an example, human beta waves are between 12.5 and 30 Hz (12.5 to 30 transitions or cycles per second), so are fast moving, and a faster staccato sound thus best represents this, while delta waves are measured at 0.5–4 Hz, so appear as more gentle waves and as such are represented in a slower, more droning, almost somnolent way.

Of course, it's not that simple; we have huge technical problems, and twelve hours before we launch a valuable new portable EEG headset is missing the vital code we need to activate it – and the only person who can sort out the problem is on holiday.

The launch

Turning off my meditation MP3, I jangle a bunch of keys until I find the piece of tape I stuck onto the gallery door key and turn the key in the lock. I am here very early on purpose because I want not only to set up for the day, but also to let the whirlwind blur of the past six months settle so I can finally appreciate that I am here. It happened.

As I flick on the lights I see the beautifully designed pop-up posters telling the story of Patient H69 that I persuaded a printing company to supply for free. They tell the story of the research project that now lies as part of this venture, and the

PATIENT H69

explanation of how we intend to extract EEG brain data from each participant. We did find a way – incredibly – to show the beautiful inner workings of my meditative brain. We have found a way to make the brain sound harmonious, and I can't wait to share it all.

There is an exquisite tension in the air even though there is no one here, and I can feel a loaded expectation already starting to simmer. This is the day I get to show the world my beach. Pushing aside the heavy curtain that conceals the room where the magic happens, I gaze over a bank of computers, screens and EEG equipment with a web of black leads strung across the walls. It looks like a giant scientific spider's web, but I have seen the technology work. I have seen what my beach looks like; I have touched my own soul, the sanctuary that resides inside the deepest part of me, and words fail me.

This gallery opens up a door to something bigger and something that might even prove beneficial on a wider scale. It was made possible because of an amazing team of people all brought together by chance with a passion for a project that had a beating heart. We didn't get the funding we applied for, but we went ahead anyway, squeezing it into weekends and evenings. My scientific learning curve has been steep, yet I am starting to understand the basic structure of our brain and the way in which it functions. After lamenting that my producing career had been wiped out by my sight loss, Jackie elbowed me with a lopsided grin and asked me who I thought had produced this.

Thinking of that moment now I hug myself with childlike glee, and dance a little jig on the creaky floorboards.

The results

The fact that we had to put on extra sessions on the last day probably goes some way to show just how impactful this small, but mighty project was. Every single slot was booked, and we often had participants queuing in case of no-shows (there was only one). Out of the 120 people who visited us, one or two were nonplussed, but the vast majority had to be coaxed out of the door at the end. But like most things in life, it's best to hear it from the horse's mouth, so here's what some of our participants said.

> One forgets that you have a brain…it gently does its things, and you pay no attention to it whatsoever; I realise that it may need some of my attention sometimes to remind me how vast and complex the science of the brain is, and how much we don't know and need to know.
>
> Wonderful. I was in a deckchair in my mind…
>
> EEG is a very boring thing, and nobody understands it – but to see your brain activity very understandably – you want to share it with the world. It was exciting taking part in this.
>
> I can't remember when I sat in a dark room and felt calm like that during the day. I'm not emotionally engaged with meditation – intellectually I get it, and now I'm wondering if I can be flexible enough to take benefit from this experience.
>
> Take it national. It's important to inspire as many people as possible – please keep going…
>
> I've actually seen a physical representation of how I think, and I really like that, but I really want to know more and analyse it. Can I come back?

The talk

Ed and I are meandering our way back to Leicester Square Tube after a night at a comedy club. This night is a double celebration as it was my gift to Ed for his birthday, but we're now also celebrating the fact that Bloomsbury want to publish my book. As we step out into the street, car horns and an onslaught of flashing neon ambush my senses, and I clutch Ed's hand tightly. This is a far cry from my producer days when a night in Soho was commonplace. We've drunk a bottle of vinegary red wine and I am regretting the choice as I fumble in my pocket to turn my phone back on.

The email from Tristan is short:

TEDx Ghent have heard about the success of The Beach
*project and asked me to do a talk on it, but I think you should
do it — it was your idea. Do you want to?*

A dream I had (for one day in the future) has just come
knocking at my door, and forgetting all about the unpalatable
wine I grin all the way home.

CHAPTER FOURTEEN

20/200

I don't think I realised until quite recently that I wasn't blind for just a few days as I initially thought, but was probably 'legally blind' for several weeks.* This should make no difference at all, but somehow that realisation lingered with me. Throughout this extraordinary experience, I have always identified myself as a sighted person who just couldn't see for a while. I never considered myself a blind person; not even for a day.

I didn't know it at the time, but during the three days it took for me to go completely blind, there would be a pivotal cut-off point that would profoundly affect my future interpretation of the world I inhabit.

As my sight slowly slid out of reach, shrinking down each hour into a smaller and smaller halo, I still recognised it. It was still the sight I had always known and used – I understood it even as it disintegrated. What I did not know was that when the metallic clicks of my internal machinery finally shut down, and as the full silence of blindness settled over me,

---

*Back in 2012 I had no knowledge of what the official measures of blindness were. I had no reason to be informed of this fact because we all hoped my blindness was temporary. However, given that I could not see anything on a Snellen eye chart until after I came home from hospital, it would seem likely that I was legally blind. Legal blindness in the USA is measured when a person's best-corrected vision is 20/200 or worse when viewing a standard Snellen eye chart. If a person sees 20/200, the smallest letter that they can see at twenty feet could be seen by a normal eye at 200 feet. I have chosen to use the American measurement as this is synonymous with the common term '20:20 vision'. The UK use metres as a measurement for visual impairment or blindness.

the act of rebooting my vision would never bring me back the same world that I had just left. What eventually returned would be a new sight, a different way of seeing and an unfamiliar vista. My experience of forty years of seeing and processing visually was to be *reset*. I have experienced something an adult is not meant to experience, let alone interrupt, analyse and then interpret.

A new normal

Ed and I always estimated six months for my recovery to be over, to be done and dusted. We discussed this hundreds of times with such utter belief that six months became the song we sang, our own modern-day intonation, sitting alongside the full-recovery chant we took home from the hospital. Ed believed this so much that he even booked tickets to see Eddie Izzard to fit with our plan. Numbers mattered – in our naivety and buoyed by optimism, we needed to have some control over what was going to happen, probably as a direct result of having had no control over what had just happened. No doctor ever gave me a timeframe, not even when nudged persistently. It didn't help that so many people asked me how long my recovery was going to take. Looking back, it's such a ridiculous question, but at the same time a completely natural one. Indeed, I narrated it into my dictaphone almost daily. *How long?*

The problem with recovery is that it is in fact just interminable waiting. I ended up bored. In truth, I'm still waiting for the drum roll and the moment I get to whip off my hat and say ta da, that's it – it's all over now. But, there hasn't quite been that centre-stage moment – yet.

Illness is punctuated by the start – time is easily delineated. But the end? Well, that can drag on. There is also a tiny part of me that still retains hope; it's hope in the adaptability of the brain, in our ability to heal ourselves and learn new ways of doing things. I don't think this is misguided, either, considering the enormous adaptation my brain has already made to accommodate my visual loss.

It's complicated

I've spent a long time trying to explain to those around me what I can see, but I'm not sure if I'm going about it the right way. Their baffled faces just scream their confusion back at me, and I find myself listening to the conversation we're not having, rather than the one we are.

Physically, I have regained pretty much the same motor function I had before – in fact, I am probably a bit more flexible thanks to the yoga I do twice a week. Yoga is my yardstick; no creeping imbalance could be hidden in a yoga class, and my physical and cognitive abilities are visible to all. Indeed, there are more than just the obvious health reasons as to why I contort myself into increasingly difficult poses week on week. I am aware that all the tiny sensors around my body have to work extremely hard to keep me off my neighbour's mat. I can tell you it's challenging – just try standing on one leg with your eyes closed and see if you don't wobble. It is only now that I understand what my body needs in order to function at its optimum level, and how it struggled to do that when I was ill.

My discovery of how my body's sensory system works explains a particularly uncomfortable moment Ed and I had during the worst of my illness. There was a time when I literally did not know if I existed. Minds are known for wandering, but this time my body had taken up residence elsewhere. I was deep underwater, floating weightless with no sense of what was up or down. Having no visual messages to orientate myself, I was drifting in a half-world that left me feeling disconnected from humanity and not of this Earth.

My question to Ed was therefore logical, and yet odd. I simply asked, 'Am I here?'

Of course, he couldn't answer anything other than yes, but I know my question disturbed him. However, it is only now that I understand that it was indeed an entirely rational question to ask. My body was accommodating huge sensory loss, so it was unable to transmit the normal data that would inform my brain as to where my body was in space. It was that very data which was telling me that I was lost.

My visual recovery, however, was a different matter. It's easier if I just say that I don't see the world like I used to. My sight didn't wholly come back — it stopped short. Complete bilateral sight loss following optic myelitis is uncommon, so there are few benchmarks against which to measure my visual homecoming.

If indeed there was a way I could accurately display what I see via some kind of biological projector, it would still not explain the way in which I understand it. Not surprisingly, I am continually asked the question, 'What can you see?' And the conversations often go something like this.

'I see most of the same basic stuff you see, but to me the world is hazy and lacking colour and light. Kind of like one of those washed out old photos.'

'Oh. So…it's blurry?'

'No, I actually have reasonable acuity — with my glasses I have 20:20 vision.'

'So, if you have 20:20 vision — you're back to normal, then?'

'No, it's as if I look through a dirty windscreen, as if there is a mistiness over everything, as if all the colour has drained away and daytime looks like dusk.'

'Hmm.'

'There's this disturbance — almost like a shimmer.'

'A shimmer? So things move?'

'Not exactly. Well, yes, sometimes things shudder to me. But we don't really know why that is. It could be something called microsaccades…'

'Oh.'

'It's complicated.'

Complicated is the word. Of course, by definition sight is subjective, so explaining one's visual experiences to another person is almost impossible. Suffice to say I normally ask my friends to wave to me when I meet them in a dimly lit pub these days, as differentiating even someone I know very well among a sea of bent heads is pretty much impossible.

Tactics

I am all too aware of some of my coping strategies these days, no doubt because I have adopted many of them as a direct result of my visual loss. Sometimes, however, these reflexes are so interwoven into the material of our lives that they become unconsciously embedded; they simply become habits. I might have overlooked some of the more subtle tactics I employ regularly had I not forced myself to stand back a little. So, hovering high above myself, I see a woman in a hat walking briskly with her head down, crossing all of the roads to avoid the dark shadows.

Mobile phones

My nimble brain is highly adaptive, and while I have acknowledged that I don't perceive the world at the same speed as do those with normal vision, I have several devices that allow me time to evaluate and assess my strategic position. Mobile phones are particularly useful for this. It is socially acceptable for me to pretend to examine my phone, while in reality I am assessing a darkened stairwell, or as in the uber-trendy Apple store in London's Covent Garden, a monstrous spiral glass stairwell with no discernible step markings. Examining my phone in this particular location had a certain paradox to it, but on many occasions my phone has quite simply – bought me time.

To others I might appear to be scrolling through emails, when in fact I am simply processing the street ahead. I might be deciphering a labyrinth of horizon lines, or where sandstone becomes concrete, demarcating the tangled up landscape. Colours are eroded, but the brightest ones still jump out in the form of road signs and billboards. Only the quiet shades lower down the spectrum remain elusive and somehow out of reach.

It never ceases to amaze me how many hazards seem to pop up in our modern lives: stairs in sports centres with no discernible edges, textured cobbles in parks and endless grey tarmac that suddenly drops away. All of these subtle (and often design) features catch me out. Harsh shadows on a bright sunny day are one of the worst offenders, plunging half of the street into a black chasm, while the other half is so bright that it makes my eyes water. I get by; I just navigate my way along. Rarely do I mention these constant blips and perils, but the chronicler in me can't help but photograph them.

If I didn't have excellent acuity I would probably be in trouble, but I can read small, high-contrast details with great ease. My brain seeks out contrast, silently beeping like a metal detector on a beach, directing my attention to any area containing dark and light together. Yes, it's safe to say that I can see road signs with a deceptive ease, and I am the one who reads restaurant menus these days, providing of course they are not printed in red. My world looks different, so in turn I look at the world differently. I have a new respect for my environment and approach it with caution. I suppose you might say I tread carefully.

People

My other great tactic is people. I was never particularly shy about asking for directions before, so this is just a natural default now. I ask for the ladies toilet in a restaurant without even trying to find the tiny logo first. I ask where the children's books are located the instant I get inside a bookshop, and if the fluorescent lighting startles my vision into a paroxysm of confusion, I engage an assistant until it settles down. In striding forwards so boldly I am in fact distracting others from the backwardness of my sight.

I live, like many others on this planet, with an invisible affliction. It is one that nobody can see, that few really understand and that most forget. I choose to live with my sight loss and accommodate it as much as is humanely possible, and in fact that in itself has created a swathe of satirical complications. I have adapted so well that on the whole few people around me have any idea of what the world looks like to me. They see a capable, confident woman with no lingering physical side effects, and as a result treat me as such. This is, of course, entirely correct, and what I have strived for, but I cannot have it both ways. I have come to realise that it is hard for others to keep remembering not only an invisible affliction, but also a perplexing one.

Ed is really the only one who understands the confusion I experience. But as I say, even he sometimes forgets. At a busy music festival he darted off to a food stall to grab a burger, and left me holding our son's hand submerged in a fast-flowing current of people. Of course, our son dashed off after his dad and I was left momentarily lost − and completely helpless. Ed had only moved ten feet away, yet I had lost both of them completely. When they returned only minutes later I was furious and frightened, not least because he had left me in the crowd, but also because it had brought home just how much independence this frustrating disability has stolen away from me. These moments are rare, but when they happen they remind us both of how our lives and roles have changed, and in turn this galvanises Ed into action. I don't think he let go of my hand for the rest of that entire festival.

Over time I have had to learn which friends to ask for help, and which social situations to put myself into. I have limits now that weren't there before, so trust has taken on new meaning. Showing my vulnerability in these situations is hard, and if I am honest, I still struggle with that. However, on the whole people are invaluable to me and my chatty persona has got me out of trouble on many an occasion. This tactic has backfired only once so far, when I was shopping in a department store some time back. Suddenly disorientated after coming out of a changing room, I made a throwaway comment to a tall woman standing nearby while I quickly assessed my whereabouts. It was her lack of response that made me do a double take, only to then realise with horror that her silence had not been rudeness, but that in fact I had been talking to a mannequin. Aside from this isolated embarrassing blunder (one I will never make again after watching Ed roll around the floor in apologetic laughter), I now make sure I ask for directions *before* I lose my orientation.

The importance of hats

I wear hats. In fact, I am quite known for the wearing of hats locally. I own quite a few, too; a dark brown trilby for autumn, a grey felt cloche hat for winter, a funky baseball cap for jeans and a variety of floppy straw hats for summer. Some are smart, some a little on the flouncy side with ribbons around the crown, and one even has buttons sown onto the brim. However, my motivation has no fashion undertones, and I can't even say that it's down to a little individual flair coming out in my forties.

As it is, hats have several downsides. They don't fit into handbags and therefore are easy to leave on trains. They tend to itch when the air is hot and humid, and this resulted one summer in the spontaneous (and instantly regrettable) purchase of a 1980s plastic visor.

No, hats are not my first choice of attire, but they have one very important attribute – a brim. My hats all serve one very vital function; they help me see more. That nineteen-year-old photography student who still lurks inside me can discern the slightest shift in contrast. Blacks become denser and more defined, leaves spring back out and the mist seems just a little less, well, misty. If I can cut down overhead glare by even a fraction, my landscape can separate out a little bit. Trees can edge themselves out of the way of each other, branches can wave and gesture me over and, most importantly, I see cars more quickly.

Hats are therefore a practical tool – a ploy. They are my millinery sleight of hand providing an invisible service. Passers-by only see the aesthetic, sometimes quirky exterior, but I know that I am in fact a hat charlatan. It's only minutiae, tiny fractions, imperceptible to a fully sighted person; but I am not a fully sighted person now, so to me it makes all the difference in the world.

Walking

I never expected that the walking which started off purely for therapeutic reasons would stick around, let alone develop into what it has become for me today.

My first few tentative steps when I got home from hospital were agony, not least because my feet felt so disconnected from my body, but also because I mourned the loss of my vitality and with it my expectations of youth. I doggedly persevered with my daily walks simply because I could not bear to lose something so important – so necessary.

Those first few hobbled steps brought with them a dissonance of noise, and I sometimes felt as if I was drowning outside. But those first steps turned into two more, and two more the day after that. Hands that had fluttered around me were calmed, and soon voices become more nonchalant and projected straight ahead, rather than at my ear.

Everyone who visited the house walked with me, all the beautiful women who put aside their free time just for me. There were the regulars visiting weekly for months on end. Then there were the busy friends who came anyway driving miles to fit me into their days. I began to realise compassion doesn't count mileage. Walking was cathartic, companionable and the way I was kind to myself, and how others were kind to me. Walking healed me from the inside out.

My walks with Genet in particular have now elbowed their way into our everyday routine; they have become part of normal life. I'm not sure what day it was when we finally talked about why we were still walking – but I know it was several years after I had recovered my mobility.

Talking is part of it, of course; and this is like mobile therapy. I like being hypnotised by nature, and the energy that seeps into me without me knowing it – sometimes I think I walk my mind as much as my body. You don't have much eye contract when you walk, so we talk about different things. We also mix it up and walk different routes, and sometimes I freak Genet out by walking on her left (which she hates), but I tease her that variety is good for her brain.

I could research the science and the psychology of why all of this is – the natural endorphins released – but I won't. This time I don't need to understand it. I am happy to let the spirituality just wash over me and for once I don't need to turn over every stone; I'm happy to just walk on top of the stones instead.

CHAPTER FIFTEEN

Time Traveller

When I was blind I had remarkably vivid dreams. The pictures were in Technicolor, fanciful and exaggerated, a toy-town world. My brain played with my memories, shuffling them while gently stroking my sleeping head as it chose another scene to project. A whole town was created, as real as I wanted it to be, so real that I could see sunlight glint on the rooftops. In my dreams I was aware of my perfect sight, and the clarity and beauty of it. It was a familiar friend holding my hand.

It made me infinitely sad to wake up, to leave that vibrant world. This merging of memory and reality was as confusing as it was frustrating, affecting not only my dreams but also my short-term memory. Even now I can recall walking down a road the day before, yet when I pull the recollection into my mind's eye to examine it again, the picture is perfect. My brain has somehow darned the holes. I was always accused of living in a dream world when I was a child. I now see that this was not a failing, but an ability I long to recapture at times when I have problems telling one day from another.

Along with these dreams, there is something I don't often talk about. It's so fleeting that it would be almost impossible to articulate if it didn't happen on a regular basis. There are times when I (almost) see again. Walking down the road, a familiar and well-trodden route, my mind flashes up an old mental image. This happens so fast that I am left with no more than a distant imprint. I just know that my brain has extracted this exact same excerpt from my mental library, and is replaying it. It's a peculiar feeling, but not unsettling. It reminds me that I know what this place should look like – that it's simply the mirror of my past.

So while I don't see the world with anything resembling *normal* sight, I can remember in perfect vision. I often cannot tell from a memory alone if an event was pre or post my optic malfunction. The wispy stills that scatter through my mind like a vintage film spool create a million fragments of time, edited and selected from seemingly random moments. Each of these frames is filled in, coloured like an animated cell, which when combined, creates colourful, vivid and *full* recall.

Flashing signs

It's easy to think, and therefore believe, that events such as the one I experienced stay with us, particularly when there is a constant reminder. I sometimes feel that I walk around with a huge sign above my head saying 'visually impaired' or 'imperfect'. However, what I now realise is that even if that placard is there, even if it is lit up with flashing red lights – other people cannot see it.

This is something you come to realise during recovery. That's why there did come a time when I felt that I had recovered enough to start living again. I wanted to join the human race and get something out of all the effort I put into getting better. I didn't have my full sight back, but I wanted to let go of the enormous breath we all took when all of this started – to just exhale. I got bored with talking about what had happened, and wanted to talk about what I was going to *do*. I decided that if my illness was going to take away my career plans, then it was going to give me something else instead.

Suggestions as to what I should do with my life were dropped casually into conversation remarkably early on. I think it's a common trait of carers to find a new focus when illness hits a family. I didn't need their advice, though; I only had to listen to my audio diary to hear what I had to do, for I had been telling myself from day one.

Perhaps naively, I thought that writing this account would be easy, a pragmatic telling of a series of events. But, maybe the raised eyebrows around me could see what effect relaying such an experience could have on someone. It's true that connecting with traumatic memories also connects with the emotions you felt at the time. I have learned that the events I went through have not left me entirely; they are still stored, sitting in a box secreted inside my inner cavities. That box is on a dusty shelf as I don't open it very often, but I live with the residue of what it contains inside.

This fascinates me more and more; my mind in its infinite brilliance cannot discern between past and present, catapulting emotions felt long ago suddenly into the here and now. When this happens I realise that I have inadvertently nudged that dusty box off its shelf, and the contents have scattered onto my living room floor. I am not immune to the lingering after effect of trauma, but I do have some ideas now on how to respond to it.

My nod to nodding off

This brings me to one thing I try not to linger on too much, but feel I should mention (in the spirit of candidness). We all have our Achilles heels in life, and mine turned out to be sleep. In my family I have always been the good sleeper, the child who crashed out on trains, planes and at sleepovers with no problem. However, when my visual system shut down, my nervous system responded with not just acute anxiety, but also insomnia. I need to be clear here – insomnia is not just sleeplessness or a bad night, nor is it tossing and turning. Insomnia is a different animal entirely. A parent at our daughter's school gently took me aside when he heard I was suffering, and offered me an ear at any time I needed it. He didn't just know first hand what not sleeping was like; he knew what not sleeping *at all* was like.

Throughout this journey I have sometimes described myself as standing on an emotional precipice. There were times when I stared down into that black hole of nothingness and felt utter despair. There were times when I tipped so far forwards that my toes were gripping the edge to hold on. My

brief but severe experience with insomnia was the one time
I fell over the edge. Spiralling down into the abyss I did a
deal with the Devil, and offered to stay blind if only he
would let me sleep. Insomnia is indeed the worst kind of
darkness.

I was lucky, though. As I fell, strong arms were there to
pull me back up, and I got the professional help I needed.
That emotional precipice is still there, but now I have to
shade my eyes to see it and I no longer fear it in the same way.
There will be others who recognise that precipice, and indeed
they may be standing a bit too close to their own right now.
I didn't find a miracle cure for my sleeplessness, but learning
that my own mind was the key to it all was a pivotal moment.
Different forms of meditation and hypnosis, plus the occasional
melatonin pill, have all been incredibly effective in helping
me manage and (most of the time) overcome this affliction.
More recently, my mind tools were put to the ultimate test.
The night before I gave my TEDx talk in Belgium I used
only meditation and breathing techniques to help me sleep.
Yet when I woke up the following morning and realised this
had worked, I was utterly taken aback. Of course, I shouldn't
have been surprised, but I still was. I will never cease to be
amazed at the power of our own minds.

Swinging the mirror around

When I got ill the cameras all turned towards me, and to
some extent my observers themselves avoided scrutiny.
However, even with a severe loss of sight I have, in my own
way, surveyed my own spectators in return. Rotating the
cameras back has offered me the true reflection of my
predicament; and exposed my carers' own individual anguish,
too. It's easy, and understandable, for the patient to often
believe that they are the only one who suffers. I know I wasn't
immune to bouts of obsessive introspection, but it is only the
benefit of hindsight that allows me to see this.

When things got really bad, the rules of engagement
changed. All interpersonal boundaries invisibly shifted as I

craved physical contact. People who I wouldn't imagine being comfortable with such intimacy dropped their reserves at the door, and dutifully curled their fingers around mine.

While of course eyes naturally transmit warmth and compassion, a cosy hand closed around mine did just the same. Touch, I now know, contains no judgement. I was intensely grateful for this physical contact, as it grounded me and stopped me from floating back off to that ephemeral planet that human beings are not meant to inhabit. It wasn't just comfort, either – they all provided a valuable source of sensory information. Plastic bags rustling into the living room piqued my curiosity, and a silent chill crept its way out of coats as they lay breathing on the sofa. Soft suede brought with it a composting outer world hidden inside its folds, and for a brief moment I would be transported into my garden. This was all valuable evidence that the world did indeed still exist on the other side of my front door. I reached out with my entire sensory army, touching the world with my ears, my nose and my fingers, absorbing whatever stimulus I could to guide me back down to Earth.

Sometimes, though, I was an unhappy sponge – soaking up the anguish that floated in behind some visitors, their silent and invisible rays seeping through my skin. This dark energy permeated me whether I liked it or not, and the culprits were almost always entirely unaware. In my heightened state it was I who heard their huffs and puffs, who read their thoughts. I think I sometimes registered their flickers of anxiety more than they did themselves. But for me, it was the hushed whispers that were the worst, often brushing the back of my neck as they crept inside my ears, for these were far more penetrating and sharp than any raised voice. What is hidden from those suffering a serious illness feels felt far more deadly than what is actually proffered.

Telling tales

Then of course there was the burning desire to share stories. This strange response to my illness was uncannily common.

A visitor would arrive and within a short period of time offered a deluge of horror stories – of mystery illnesses, and quite often, terrible outcomes. I got quite practised in recognising the big intake of breath, their torso inching forwards; and I knew a Terrible Story was about to be purged.

While I am a lover of real-life tales, these chilling tomes of doom often left me with irrational fears ricocheting inside my head for days on end. I was obviously in a particularly sensitive state for quite some time, so it didn't take much to trigger my fear. It's common behaviour for those around us to empathise with someone they care about; in fact in psychology it is recognised as 'downward comparison'. In some ways they are attempting to portray a more horrible situation than your own in order to reassure you that there are others more worse off than yourself. However, while of course that is often true, one vital fact remained – *I didn't care.*

I didn't care because I couldn't. I didn't have emotional space for anyone else's drama, as I was concentrating so hard on riding my own rollercoaster and not falling off.

There was a time when, during my recovery, I was introduced to a relatively new neighbour. Within minutes of sitting down with her coffee cupped in her hands, she opened her mouth with *the look*. This time, however, I had to interrupt.

'You're going to tell me a horror story about some awful illness someone you know has had?' Her mouth remained open, but she sat back with a bump. 'Don't worry, everyone tells me horror stories, except that I really don't want to hear them. It's some strange reflex, but I'd rather you didn't today.'

Her shock dissolved into a chuckle and I think I can safely say that was the moment we became friends. During my recovery honesty took on a whole new meaning, as ultimately for me it became a form of self-preservation. Thankfully, in this case diverting from the conventional rules of conversation was accepted with very good grace.

We all have coping strategies, and it is only fair to acknowledge the coping mechanisms of those around me, as well as my own. In fact, I adopted parallel cognitive strategies in some ways, often telling myself that other people don't recover from blindness and I am doing so. I truly appreciated the milestones I hit during my recovery, and championed each small achievement. These hurdles were to become the foundation of my internal rhetoric. It's an interesting consideration, then, to see this collision of coping mechanisms, so misaligned, yet so well intentioned. It was comedic and somewhat ironic, too, to find myself three years later listening to a friend telling me about her distressing break-up, only to find myself leaning forwards and taking in a large breath.

Rainbows

Trauma doesn't hit just one person; it hits a family and becomes a shared burden. I often wonder if my illness left any long-term impact on Ed or our children; if they too have their own invisible scars. In the first weeks of my return home my daughter regularly dissolved into paroxysms of anger, particularly at night-time. I have no way of knowing if her brain was responding to the tension and fear that you could at times almost taste in the air – fear that I am sure emanated from me. As a child that craved routine, it was her life that had been turned upside down, too.

Throughout my illness Ed shouldered a huge burden. He had no idea if I would recover, and if I did how disabled I might be. But even in the darkest times Ed was there, and even when I couldn't see him I could feel him next to me.

My illness has, I think, allowed my family and I to all understand each other just a little bit better. Nowadays the children subside, climb back down the curtains and curl around me when it is anything to do with Mummy's eyes. They know I have lost rainbows, and that's serious. I dread those rainy, yet bright days when they run to the window and prod the glass with their fingers. Cries of 'rainbow' punch me in the stomach as I reluctantly follow them to stare out at the sky. If the colours are faint the chances are that it won't be visible to me – it needs has to be a bold, smiley rainbow for me to see it now. My son knows this all too well, so as we stand there looking out he slides his hand into mine – just in case.

My sight loss is also a device for my children – something to be used if an item is missing, or forgotten on the school run. A mishap is often blamed on my sight. Sometimes they are remarkably sensitive about what I can see, and birthday cards are painted in bright bold colours so Mummy can see them. They intuitively understand; they *get it*, and I wonder if it will make them a little more perceptive and thoughtful as they move into adulthood.

Children adapt, and they are remarkably resilient. The first conversation I had with my five-year-old daughter after having not seen her or spoken to her for fourteen days was one I will never forget; and one I cannot possibly repeat. Suffice to say, we talked about butterflies, and I still have the one she sent me here in my pocket right now.

Perception

Seeing is believing. This idiom is clearly provocative, but if one reads it a little too quickly some of the subtlety could be missed. Indeed, how you read it could even offer you some insights into how your own brain processes information. For me, these words resonate with my own visual recovery, and my renewed understanding of the world.

Of course, this can very quickly just become semantics. When any discussion regarding my returning sight came up, it was talked about in simple terms – more light sensitivity, acuity, seeing colour again. We didn't discuss how I might interpret those visual messages. Until I lost my sight, seeing had been an unconscious act, a valuable gift. This is not an exaggeration – I have always been utterly

absorbed by my visual world, which is probably why this
book now exists.

Seeing is one of the most complex functions our brain
constantly performs. There have been times when my
journey morphed into a fantasy world, and I can recall
moments when I honestly could not believe what I was
seeing, yet I had to believe it, as it was what my brain was
telling me was true. I have seen what I needed to see in order
to recover.

Those inconspicuous small details that I was seeking out
on Mowbray Road were what turned that street into my own
personal science lab. I've learned that this is basically how our
attention works. Vision is one thing; being attentive to what
is in front of us is another thing entirely. If we are occupied
by one thing, it's easy to miss other things that are (to coin a
phrase) right in front of our noses.

The more I have learned, the more these individual
perceptual quirks fascinate me. However, I have to admit
to a frustration that I have no choice in the limited picture
I see – no matter how hard I strain. I can't help but feel that
those around me have unrestricted access to so many
magical moments that they unconsciously choose to ignore.
Of course, it is entirely unreasonable for me to want them
to see what I would see (if I could) and to disregard what
they do see as unimportant. Yet whenever a distracted
commuter barges past me, I fleetingly wonder whether if I
borrowed his vision for just one minute to once more suck
up this magnificent world – would he even notice that it
had gone?

My response to my illness was unique and part of the
interwoven neural network that had already made me into
the person I am. But we don't stop and see, we just see – we
look through the window every day, yet we don't normally
notice the glass. While I haven't had a chance yet to unravel
every strange visual occurrence during my illness, what I
have uncovered has opened my eyes, in every sense of
the word.

Not the end

My illness was sudden and frightening and it stopped me in my tracks – for a while. However, how an illness changes you is actually your choice. It's not what happens that matters so much as how you choose to respond to it. We do have choices, lots of them if you hunt them out – although that can be the tricky bit. But, the way I see it is that if those choices are not obvious, then you just have to carve them out for yourself. One of my inspirations for writing this book was Jean-Dominique Bauby – author of the stunning memoir *The Diving Bell and the Butterfly*, which he wrote in 1995 while suffering from locked-in syndrome following a stroke. This left him unable to move his body or speak, so he dictated his memoir letter by letter by blinking his left eyelid. I believe he was a man who from the outside looked as if he had the least amount of choices left, yet he went ahead and did the impossible anyway.

One of the most unexpected bonuses of this episode in my life was that the scientists and clinicians who guided me to this point would in time become my friends. I hadn't considered what they might gain, so it's a humbling revelation that my patient's account has value to them. When I look back there were so many things thwarting our EEG project, yet it happened anyway. Being part of a project like that connects you in a different way. Observing strangers so deeply moved when they saw and heard the workings of their own meditative brains was a privilege I won't forget, and I hope one day we get to repeat the experience.

Researching my illness has at times been a healing process, but on the whole it's been fascinating. Of course I'm still no scientist, but I am definitely a science-geek now. I realise that in the process of dissecting my story I have learned a huge amount about myself, but mostly I've learned that it was the science that kept me sane.

Back in a time when Ed and I had been seeing each other for only a few weeks, he met Jackie and I at the pub one night. Feeling a little mischievous and thinking I was out of earshot, Jackie nudged him and asked how he felt things were going between him and I. Without missing a beat, Ed picked up his beer and, taking a sip, quietly replied, 'Fine, but ask me again in forty years.'

I feel rather the same way when people ask me how things are for me now (but perhaps I won't put quite such a long time frame on it), so I'll leave you with Ed's words, not mine – my husband and supporter who I will never see eye to eye with on colour, but who is always there. This time I'm going to let him have the last word. It doesn't happen often.

Appendix I

The beach: A visualised sanctuary

There is a delicious coolness that can only be captured by wiggling your toes deep down into fine sand. It's like the cool underside of your pillow, which on a hot, stuffy night you sleepily flip over. These things are unbelievably satisfying and somehow even nurturing, yet so often only experienced on the very edge of our consciousness.

I can feel the prick of heat on my skin but I am not worried as it's too early in the day yet for the sun to do me any harm. Waves slop and break leisurely on the sand in keeping with this slow part of the day. The sea is the most azure blue I have ever seen, and white light sparkles and dances on the surface. In a flash I dive straight in, as sleek and glassy as a dolphin, my eyes glistening orbs. My arms merge into my body as I glide effortlessly under the water, all fears and contact-lens concerns left in the outside world. My amphibious form sends ripples and wakes as I cut through the water, energy, light and joy speeding me on.

The colours around me are fantastic, hyper-real and my kind of beautiful. My brain happily supplies boxes of memories for me to riffle through, searching out the brightest blues, the most ochre yellows with which to paint my picture. I have created the most evocative and safe sanctuary, celestial in its meaning and almost as real as any place my physical body might visit.

In this wonderfully calm place my memories allow me to see the pale ice-blue of my children's eyes and the candy-floss pink of their tongues. Unlocking these spiritual treasures, I am nourished and life's rainbow is once more part of my life.

I am called back, and find myself in our living room on a Sunday morning. The part of my mind that delivered me back drifts easily away smiling and serene, while my eyes blink rapidly and my physical body gingerly stretches out one stiff leg at a time.

The azure sea is replaced with the usual grainy voile and muted landscape I now know so well. It has become my 'new normal'. My sense of sadness and loss is always quite acute on my return. I am thankful for my sanctuary, for the haven it offers me, the opportunity to tap and extract the millions of stored images filed away in my brain. I have created a collage of frozen moments and glued them all together into a landscape I can visit anytime; and believe me I have been here so many times that I should probably have an annual pass. Meditation isn't just something I do; it's somewhere I go.

A conversation

There have been many times while writing this book when words have failed me, or perhaps I failed with them. One of the biggest challenges I faced was to describe what I came to call 'the beach'. The first problem is that the beach doesn't actually exist. This didn't give me a great place to start, for it means that any willing reader will need to walk with me down the dark neural corridors of my brain to reach this wonderful mental place. Dr Ann-Marie Golden, who worked with me on *The Beach* EEG project, agreed to do just that, not as just a psychologist, but as a friend. This is what happened.

H69 So, where do we start?

Dr AM I think we can start with what happened. Then we can go to why and how it came about. You know, I was intrigued when you first described the beach to us before the project even took off. It sounded like a complex coping strategy that proved to be very beneficial for you, at a time when you were suffering acute anxiety. So, it's probably worth starting with what you considered the beach to be – for you personally.

H69 That's interesting – the ownership thing. I think I have to start with that actually. It has to be yours. When Tristan asked me to give a written description of my beach after I had the EEG session I procrastinated for ages. Even writing out what happened inside my head was so exposing. It was awful – I felt as if I was waving my knickers above my head! Of course, I got the logic of this because that way he could work out exactly which section of the data related to the beach itself.

Dr AM Would you describe it as a sense of nakedness? A sense of raw vulnerability?

H69 Yes, I felt very exposed – because I felt as though he might find out something that even I didn't know about myself.

Dr AM This is bringing in vulnerability, and the courage to be vulnerable. This is important, and while it's less to do with the beach itself, it's an important part of your processes. Can you describe your beach to *me*?

H69 Okay, it's a beach; it's that simple, really. It's a made-up place inside my mind. I imagine it. But it's always in full colour and I can see lots of detail, and I control everything that goes on there. But – even as I talk now this whole thing has a reality it was never meant to have, from the project and talking about it so much. Describing it in words always seems so wrong. You have to feel it.

Dr AM I understand that.

H69 Part of my mind that needed the beach for protection feels as if I'm giving away something. As if I'm admitting to a fantasy dream I'd had about someone, as if there are implications by talking about it, even though I don't know what they are.

Dr AM Yes, and this is all understandable; it was an important place for you. However, even when you describe it, I still can't see it in the way you can. Those images will always be yours. Your beach won't be the same as my beach. Most people are capable of visualising and creating their own safe place. In fact it doesn't need to exist for others; what needs to exist for others is some small amount of imagination and creativity.

H69 Yes, that's so true, anyone can do this. I've missed out a crucial thing actually – how I got to the beach, because there was kind of a process. In the hospital it all started with me using the breathing – it neutralised what was going on. I would use golden thread breath – you know, breathing in for a count, then out for a longer count. It's visual so I loved it and it distracted my brain. Somehow, after that, I don't really know when, I started creating a sanctuary, and it was this beach with soft golden sand and blue skies, and it started to feel real somehow – physical. It was made up but I kind of knew it in some weird way; it had a familiarity to it, but parts of it were definitely invented. I would sink down inside myself and mentally transport myself there, and would instantly feel calm and safe.

Dr AM Okay, so picking up on the breathing bit – on a brain level, as far as physiology is concerned, what is happening is that when we breathe in we get oxygen, then this gets turned into carbon dioxide, and if we don't breathe out regularly or for long enough, we're essentially holding back carbon dioxide, and the brain panics. It says, 'I can't cope with this!' So what you're doing with that type of breathing is releasing any carbon dioxide. From a physiological point of view, it's no wonder that it calmed you down.

H69 Sometimes I made my breathing loud, really audible, so I could somehow drown out the bad thoughts in my mind. It was as though my brain couldn't concentrate on negative thoughts and this loud breathing all at the same time. It was a brilliant distraction.

Dr AM Okay, so the main job of our autonomic nervous system (ANS),* is to observe the internal environment and

---

* The ANS is the part of our nervous system that controls our bodily functions without our conscious control – like breathing, the beating of our hearts and our digestive system.

also regulate it. Autonomic just means independent from our conscious mind. If you think of the ANS as a team of animals – they will always follow the leader, and the breath is only part of the ANS that is consciously controllable. So the lead animal starts the breath and the rest of the team follows.

H69 I like that.

Dr AM The breath is really important in all of this – it's talked about in mindfulness, but I don't know if we ever really say why. Breathing in delivers oxygen and relevant nutrients to every organ, including the brain and our musculoskeletal system. Without the parasympathetic component of the ANS,* which balances everything, the systems can get overwhelmed and you might then feel confusion, dizziness, spacey-ness, fear, anxiety and other forms of distress and hyperarousal. Therefore, breathing poorly will result in blocked or excessive swings of energy. The breath is key.

H69 That makes complete sense. What about the distraction thing, though?

Dr AM This is essentially mindfulness in practice... a good example is distracted drivers who, as well as driving and paying attention to the road, are also thinking about where they are going – maybe talking on the phone – and the GPS might be telling them where to go and there might be some children in the back of the car....you know? For years we believed that we're good at multitasking, but it turns out that's a myth. The brain can perform two tasks at the same time but handles them in sequences, switching between one

---

* The parasympathetic system is responsible for what is known as 'feeding and breeding' activities while the body is resting. This includes digestion, sexual arousal, salivation, lacrimation (tears) and urination. This system is complementary to the sympathetic nervous system, which is sometimes known as the fight-or-flight response, occurring when we're hyper because we fear an attack or a threat to survival.

task and another – which can be very tiring. Obviously, our brains can juggle tasks really quickly and this makes us believe that we are multitasking. So, a distracted brain is essentially just a process of attention switching. So tell me, why do you think you created this beach for yourself?

H69 Hmm. I don't know really, but I just did – I do know it was more useful than anything else I had as a resource at that time. It was also completely controllable by me – that was the best bit. As a patient you're powerless much of the time, yet I had something I completely controlled. Particularly as I couldn't move or see; it was something I could flick on or off, and it had speed – I could get there quickly. There was something about it being on tap and uncomplicated and easy. It was instantly available when I wanted it; I didn't have to ask anyone else if I could visit it and didn't need anyone else's permission. There wasn't even a dialogue about it. This is the crazy thing – at the time I didn't tell anyone about it. It was such a reflex, such a natural thing – I almost wasn't aware of it. I know I used to say, 'I'm going away for a bit,' but I don't remember anyone ever asking me where I was going. So – what do you think I was doing? Why did I create it?

Dr AM The beach kicked in as part of your survival instinct. Fear provoked it. Survival is survival of the fittest, and it's instinctive. In the world of psychology it's understood that we were animals before we were humans – so we had to survive and fend for ourselves. Survival is a strength we humans have.

H69 So what makes our survival instinct kick in, then?

Dr AM If you think of dividing your brain – we experience the fear response in the amygdala, which is the part of the brain that deals with emotional responses like fear. This is basically the hub of our survival instinct. Essentially in evolutionary terms we have had to survive and reproduce – this is our essential purpose, and why we behave and interact with others. Kahneman wrote about it in 2002 and he won the Nobel prize in economics.* He talked about how we make decisions; we don't always get the best outcomes, but they are most efficient in terms of time and efficiency – they get us by. He discussed hot emotions like surprise and disgust, which we instantly experience and which are very powerful and imminent to our survival. We experience these emotions when we come across, for example, sour or rotten food, or attack – and they increase our chances of survival. The fear you had of being attacked by the darkness made you respond in the way you did. But you used other parts of your brain, too. What you did was to tap into that part of your brain – the 'being present' part and 'self-compassionate' part – and that potentially created your new neural pathways.

H69 That's pretty mind-blowing. Does that mean my mind actually created more brain cells?

Dr AM Yes, it's getting really exciting actually. Dr Richard Davidson at the University of Winsconsin-Madison has done research into how we can train our minds and improve our performance. One of his studies has shown that after one day of intensive practice of mindfulness, alteration in gene expression – in other words our DNA – can be detected. His finding shows that meditation reduces the expression of genes that promote inflammation. This is a very revolutionary finding if you consider that inflammation is implicated in almost everything – from heart disease and cancer, to arthritis and what you suffered...

H69 So in a weird way I was kind of helping myself on a cellular level?

*Daniel Kahneman is an Israeli-American psychologist who is well known for his work on the psychology of judgement and decision making, as well as behavioural economics, for which he was awarded the Nobel Memorial Prize in Economic Sciences (shared with Vernon L. Smith) in 2002. His empirical findings challenge the assumption of human rationality prevailing in modern economic theory.

Dr AM Yes, in essence you were. The beach was about survival but also a refuge. It was about going somewhere and not being in the present. We would call it conscious avoidance, in fact. You were consciously avoiding the horror of your immediate situation.

H69 I've sometimes felt a little guilty, or that I was self-indulgent about going to the beach so much. Other people didn't understand what I was doing, let's face it. It was very public and we don't normally meditate in such an obvious way with people around us not understanding, yet watching us. I had the feeling sometimes as my family whispered around me that they thought I was dozing, but I wasn't. I was present, but in an entirely different way.

Dr AM You were protecting yourself, and what is interesting is that you did it without any prompting. Mindfulness and self-compassion work means that when we take care of ourselves we activate a 'self-soothing' part of the brain, as opposed to 'giving into the fear' that our mind is creating. This stops us from being taken over by anxiety; it's not that anxiety doesn't happen, but they can happen in parallel. This is connected to our fight-or-flight response. What you were doing at the beach is *both* fight and flight. You did both at the same time. Running away – but to a safe place – you retrained yourself to survive for living in the darkness. There is an element of avoidance, because the reality was so awful; it was pure self-preservation. But self-preservation is a survival instinct. It's interchangeable – fight and flight come into play when we perceive a threat to our existence, and you losing sight *is* a threat to your existence. Your sympathetic nervous system activates rapid emotional psychological and physical changes.

H69 It is interesting that it's both fight and flight, but you know I had never heard of that phrase before all of this!

Dr AM Well, it's what you did. It kicked in because emotionally you could feel fear, and at such a time the senses are heightened and you make faster decisions – you got a shot of adrenaline. You didn't sit and deliberate about what you should do. You made a cerebral natural response, which in essence is tied to self-preservation.

H69 I know I used all sorts of mind tools. I tapped into the best feelings I'd ever had in my life to sort of piggyback them. I used my memory to give me good feelings to replicate them again – almost so I could remember what that felt like. Is this rewiring?

Dr AM Yes. What we know is that when we are mindful our neurons are firing and they are rewiring. The old stuff is dying and potentially the synapses are there and can be refired – recreated or as you say rewired. I think that is an important aspect of what you did. This is basically neuroplasticity. It occurs inside us every day when we have new experiences. The more often neural pathways fire, the stronger the connection will become.

H69 The more I learn about neuroplasticity, the more interested I get. It's just so part of our lives, isn't it?

Dr AM Yes – there is so much on it now. The Canadian psychologist Donald Hebb says 'neurons that fire together wire together'. So he's basically saying here that our brains change based on our experiences. Then, by our experiences changing, we can actively reshape our brains. So when we are learning mindfulness, our ability to intentionally be aware of our experiences as they are unfolding, we develop a capacity to pause before we react. Daniel Siegal calls this 'response flexibility'.

H69 I really want to get my head around it all. Perhaps this explains why I felt my beach was addictive at times?

Dr AM It's possible that by replaying these scenarios and going to the beach as often as you did you felt safe and connected. In essence you were creating new connections. The addictive part also makes sense since it is a place that is offering you solace, belonging, safety and connectivity. This is absolutely neuroplasticity at work...

H69 I wanted to ask – do you suppose I responded to a trauma in the way I did because I already had experience of meditation and self-hypnosis? Is that a really obvious question?

Dr AM The answer to why you did what you did is that it was your own unique response. Everyone responds differently in these situations. Yes, in your case you already had some resources, didn't you?

H69 Yes, I guess if you put ten people in the same place as me they wouldn't all meditate – they'd respond in their own ways. So why did I do that? Actually I do know why. It was because of a hypnobirthing experience, and I'd had quite a lot of meditation exposure, but I don't think I'd quite had the eureka moment, that 'wow, I'm doing it right', because I don't think that anyone ever has that. I think meditation offered me a lot of interest – but the thought of doing it every day was crazy! So I didn't have that click then, but the self-hypnosis was different. I did that religiously when I was pregnant the second time around for one simple reason – to fight fear, because I knew that it was possible to rewire your brain, so I had quite a methodical process. I had an aim, a mission, and I had a set of tools and homework, so I did it religiously every night. I never knew if it had worked until I had my son and it was put to the test. So here I was with no safety net and no support and it was fine. It was only hindsight that allowed me to see that afterwards.

Dr AM Here we are back to how you knew what to do – and in fact you were prepared for it. Different people might respond differently based on their coping skills and circumstances. We know there are a number of risk factors that increase people's susceptibility to emotional and psychological trauma, such as an accident, injury or attack, or workload, or suffering from an illness or bereavement. Experiencing previous traumatic events and even childhood trauma can increase the risk of future trauma. But survival instincts are innate in all of us and this is the time when we revisit what resources we have at that moment, assessing what is happening to us, what we are doing and what else we could be doing.

H69 I see now how my brain picked out all of the resources I had. With hindsight it was a very powerful thing to give up all control, to let go and just to trust myself. I allowed the biochemistry in my body to take over. I suppose what my body did was to find the polar opposite place to where I was in reality. I don't know where the beach came from. It's all a bit muddy now, but I know part of this was that I wanted to practise seeing when I was on the beach. That sounds mad now I've said it out loud!

Dr AM It's not mad at all when you consider that different parts of our bodies, like our vision, experience more strength and stamina when we're visualising. It's something we start to work with from a young age, using our imagination – so that's what you were doing. To bring in the Pavlovian response here (which is classical conditioning), by going back to the beach all the time, you essentially conditioned yourself to keep on seeing – to keep on activating your optical muscles so as to keep on seeing. You were conditioning the visual part of your brain every time you revisited the beach. You knew some of the theory of atrophy, so you responded to that knowledge.

H69 So creating a visual sanctuary was what I was always going to do – because it needed my history, my life and my experiences to build it.

Dr AM Yes, you used layers of experiences to create it – a hybrid of hypnobirthing, golden thread breath, meditation, memories and coaching techniques. You felt in control and had a box of tools, a number of skills that you adapted, and you said this was where you would start – and let's see where it took you. It probably did work to an extent, but you thought about what else you could add – let's try X, Y, Z, the more the merrier, there is nothing to lose here. By doing this you were retraining the muscles and firing the neurons in the brain, and saying, 'Hello, I exist and at least I can see the beach!' And then we go back to 'neurons that fire together wire together'.

H69 I guess we all use what we have. Any sanctuary is unique in your own right, and it's your own coping mechanisms. It's a relief that there doesn't have to be a template, a standard way of doing it. When we did the EEG project lots of people said that I was doing it wrong. But one of the things for me is that you are unique – your coping strategy is going to be all the more powerful for its uniqueness. Okay, there are guidelines to follow if you need a rigid template, but if you use your own brain in your own way, you can find your own thing.

Dr AM I think what is relevant here is Jon Kabat-Zinn's seven factors that are necessary in the practice of mindfulness: non-judging, patience, beginner's mind, trust, non-striving, acceptance and letting go. I think without knowing it you applied these attitudes as there was nothing else to do or work with. It's important to know that none of us is an expert in this. I tell my clients all the time that I forget to use mindfulness, even though I know it can help me enormously and even though it got me through my cancer treatments. I still don't use it all the time and have to remind myself to do it. It's the way it is; we are all beginners and we are all human.

H69 That is so true – and it's really helpful and such a relief to know that. I told many people at the exhibition to give themselves, and their brains, a break. We are so hard on ourselves.

Dr AM We are, but we have mindfulness and self-compassion if we are prepared to allow ourselves to just be present, kind to ourselves and curious about being here – right now. It's our choice.

H69 Absolutely – curiosity is what sparked all of this!

Appendix II

Visualisation guide
patienth69.com/meditation

If, after reading this book, you find yourself curious to know more, or even want to try out visualisation for yourself, I have something for you. To save you the trouble of Googling one, Lisabetta Vilela, my meditation teacher, and I have created a simple guided meditation MP3 that you can download for free. Michael (my composer friend) recorded it for us, and even created a meditative soundtrack for it. Consider this the first step if you've never tried visualised meditation before, or if you have, this is another one to add to your collection. Enjoy.

Further Reading

For those seeking some additional reading, this selection only skims the surface. I've included scientific papers (you may need to pay to view some of them), articles and books I've mentioned in the text.

NMO and NMOSD

Marrie, R. A. & Gryba, C. 2013. The incidence and prevalence of neuromyelitis optica: a systematic review. *International Journal of MS Care* 15(3), 113–118.

Vision

Barry, S. R. 2009. *Fixing My Gaze: A Scientist's Journey into Seeing in Three Dimensions*. Basic Books, New York.

I've read most of Sach's books now, but this one started my journey into neuroscience:
Sachs, O. 2011. *The Island of the Colour-blind*. Picador, England.
Hull, J. M. 2013. *Touching the Rock: An Experience of Blindness*. SPCK Publishing, England.

More reading on how our brains can predict what we see:
Clark, A. 2013. Whatever next? Predictive brains, situated agents, and the future of cognitive science. *Behavioral and Brain Sciences* 36(03), 181–204.

Visual acuity and contrast sensitivity:
Bodis-Wollner, I. 1972. Visual acuity and contrast sensitivity in patients with cerebral lesions. *Science* 178(4062), 769–771.

Primary reference for Hubel and Wiesel:
Hubel, D. H. & Wiesel, T. N. 1962. Receptive fields, binocular interaction and functional architecture in the cat's visual cortex. *The Journal of Physiology* 160, 106–154.

Study on patients blinded by cataracts in childhood, but who regained vision later in life:

Kalia, A., Lesmes, L. A., Dorr, M., Gandhi, T., Chatterjee, G., Ganesh, S., et al. 2014. Development of pattern vision following early and extended blindness. *Proceedings of the National Academy of Sciences of the United States of America.* http://doi.org/10.1073/pnas.1311041111

Opponent process theory of colour:

Hurvich, L. & Jameson, D. 1957. An opponent-process theory of color vision. *Psychological Review* 64, 384–404. doi: 10.1037/h0041403

Synaesthesia

Ward, J. 2008. *The Frog Who Croaked Blue: Synesthesia and the Mixing of the Senses.* Routledge, London and New York.

Ward, J. 2013. Synesthesia. *Annual Review of Psychology*, 64, 49–75.

Sensory loss

Blanke. O. 2012. Multisensory brain mechanisms of bodily self-consciousness, *Nature Reviews Neuroscience* 13, 556–571 (August 2012). doi:10.1038/nrn3292

Well-being

Insightful article surrounding neuroscientist Richard Davidson's approach to self-compassion work and how we can train our brains to assist our own well-being:
www.mindful.org/science-reveals-well-skill

Summarises the research of the Nobel Memorial Prize winner (Economics) around the themes of cognitive biases, which came up during my talk with Dr Golden:

Kahneman, D. 2011. *Thinking Fast and Slow.* Farrar, Straus and Giroux, USA.

More about Donald O. Hebb, the neuropsychologist and his theory of Hebbian learning that resulted in the phrase 'neurons that fire together wire together':
https://en.wikipedia.org/wiki/Donald_O._Hebb

Meditation

Kabat-Zinn J. 2013. *Full Catastrophe Living: How to Cope with Stress, Pain and Illness Using Mindfulness Meditation.* Piatkus, UK.

Just inspiring

Bauby, J. 2008. *The Diving Bell and the Butterfly.* Harper Perennial, London.

Finally, check out TED.com if you haven't already. Some great talks on here (if you want to find mine it's Vanessa Potter, 'The Art of your Mind').

Acknowledgements

First and foremost I have to thank my family and friends for wrapping me up in a blanket of love and looking after me when I was ill. I am indebted to you, and this book only now exists because you were there for me then.

Writing this account has required me to drink gallons of tea in the many hospitable cafes of Crystal Palace. It has also required the inexhaustible patience and expert guidance of a number of scientists and clinicians. I say patience, as I am sure some of my earlier questions were, well, quite frankly – nonsensical. Thanks not only for your answers, but also for educating me on the science of myself. In no particular order, my thanks go to Dr Tristan Bekinschtein, Dr Will Harrison, Dr Ann-Marie Golden, Barbara Jachs, Professor Jamie Ward, Dr Giles Hamilton-Fletcher, Dr Noelle Blake, Dr Emiliano Merlo, Marisa Rodriguez-Carmona, Ruth Perrot and Nikki Adams. Also Dr George Tackley, Kay Day and the NMO team at the John Radcliffe Hospital in Oxford.

The Beach EEG project would never have happened without the absolute dedication of Lizzy Moriarty, Tristan's lab, Andres Canales, Srivas Chennu, Sridhar Jagannathan, Michael Powell, Jackie Ankelen, Barbara, Ray Christodoulou, Deirdre Janson-Smith, Susie Fisher, and Dan Shorten and his student Dominic Baker at the Guildhall School of Music and Drama. Thanks also to the Cambridge Science Festival organisers and the participants themselves, and all of the invisible people behind the scenes who made that project so amazing.

Thank you Lisabetta and Michael for the wonderful meditation guide (a freebie download in Appendix ii).

I'd also like to thank Dr Fred Schon (*The Big Man*) and his team at St George's Hospital for looking after me.

I can't forget Ceri Levy for the introduction, Jim Martin at Bloomsbury for believing that a blog could be a book, Anna MacDiarmid for her tireless editing (and answering of questions), and everyone who read (and reread) this until I got it right.

Not least I am grateful to Mum for always being there when I need it, and for the support of my children and my rock – Ed. I don't need to say more, do I?

Patienth69.com

Index

To come